高职高专计算机教学改革 新体系 教材

基于华为eNSP网络设备高级应用实验教程

李 锋 编著

清华大学出版社
北京

内 容 简 介

本书是基于华为 HCIP 认证考试大纲和 eNSP 模拟平台，参考国内外有关文献资料，结合编著者多年教学经验而编著。在内容安排上遵循循序渐进的原则，注重基础性和实用性，精选了 14 个企业工作任务，对理论知识和核心技能加以层层深入的讲解。实验项目基于华为 eNSP 模拟器，内容涉及 MSTP 多生成树协议、VRRP 协议配置、基于 MSTP＋VRRP 实现网关冗余和负载均衡、OSPF 身份验证和虚链路、路由引入、IS-IS 路由协议、BGP 路由协议和路径选择等。

本书采用"基于工作任务"的方式组织教学过程，对理论知识和核心技能加以层层剖析，目的是学以致用。为便于读者自学，本书对配置脚本做了详细分析并提供了笔记，既可作为高职计算机网络专业的配套教材和 HCIP、CCNP 等认证教材，也可作为初学者的入门参考书。

本书配套课件、实验录像、在线实验和讨论答疑网络课程站点，该站点于 2016 年荣获第十五届全国多媒体教育软件大赛二等奖（教育部指导、中央电化教育馆主办），遴选为广东省在线开放课程（广东省教育厅）；2019 年获得第八届全国高等学校计算机课件大赛二等奖（教育部高等学校计算机科学与技术教学指导委员会）。

图书在版编目(CIP)数据

基于华为 eNSP 网络设备高级应用实验教程/李锋编著. —北京：清华大学出版社，2023.7(2025.1重印)
高职高专计算机教学改革新体系教材
ISBN 978-7-302-63802-5

Ⅰ. ①基…　Ⅱ. ①李…　Ⅲ. ①计算机网络－高等职业教育－教材　Ⅳ. ①TP393

中国国家版本馆 CIP 数据核字(2023)第 105793 号

责任编辑：颜廷芳
封面设计：常雪影
责任校对：袁　芳
责任印制：宋　林

出版发行：清华大学出版社
网　　址：https://www.tup.com.cn, https://www.wqxuetang.com
地　　址：北京清华大学学研大厦 A 座　　**邮　　编**：100084
社 总 机：010-83470000　　**邮　　购**：010-62786544
投稿与读者服务：010-62776969，c-service@tup.tsinghua.edu.cn
质量反馈：010-62772015，zhiliang@tup.tsinghua.edu.cn
课件下载：https://www.tup.com.cn，010-83470410
印 装 者：三河市君旺印务有限公司
经　　销：全国新华书店
开　　本：185mm×260mm　　**印　　张**：9.5　　**字　　数**：240 千字
版　　次：2023 年 9 月第 1 版　　**印　　次**：2025 年 1 月第 2 次印刷
定　　价：39.00 元

产品编号：102503-01

前言

FOREWORD

党的二十大报告指出，统筹职业教育、高等教育、继续教育协同创新，推进普职融通、产教融合、科教融汇，优化职业教育类型定位。做好教材的高质量出版工作是办好职业教育的重点之一。基于此，编著本书。通过本书的学习，学生加深了对网络理论知识之间逻辑关系的理解，对华为认证体系在纵向和横向上有了一定的认知。

但是，由于学校的教学内容总是滞后于企业技术的进步，因此学生实践能力与企业需求之间会存在错位，究其原因，首先，由于学校经费不足，实验场所有限，设备器材陈旧，实践教学难以贴近社会、贴近企业，学生难以做到学以致用；其次，在传统实践教学中，学生在实践课程中往往扮演的不是实验主体的角色，而仅仅停留在教师演示和理论讲解的层面，只是对实验讲义操作步骤的简单模仿，再加上时间和空间上的限制，课程实践模式简单，方法单一，过程固定，学生创新思维和动手能力并没有得到真实有效的培养和提高。

为解决这一教学难题，编著者在课程授课过程中将虚拟实践环节引入课堂，精选企业典型工作任务，撰写了配套教材，利用华为 eNSP 软件模拟网络拓扑，选型仿真设备，将虚拟实践方式和传统实验教学结合起来，着重培养学生动手能力，有助于提高对华为认证体系的横向和纵向认知。

在具体实践教学过程中，宏观上采用基于工作过程的任务导向教学法，微观上综合运用“角色扮演”“虚实并举，软硬结合”教学方法引导实践内容由浅入深逐步展开。

本书的特点如下所述。

（1）采用“角色扮演”实践教学法提高学生参与热情。

“角色扮演”实践教学法是以学生为中心，通过团队合作、教学互动、师生互动以提高学生参与积极性的教学方法。在具体教学实施过程中结合企业典型案例，由教师充当竞标企业领导，企业下设网络规划工程师、设备配置工程师、网络调试工程师、标书编写与投标人员等多种岗位，让学生根据岗位职责交替选择不同角色，投入具体的工作环境和工程任务中，通过分组讨论、团队协作，演绎一场竞争激烈的竞标过程，共同完成复杂繁重的网络工程，从而培养学生综合分析问题和解决问题的能力，具备吃苦耐劳的职业素养、严谨科学的处事方法和积极向上的生活态度。

（2）通过“虚拟仿真、实物结合、实践验证”三步学习法启发学生思考，提高认知水平。

对网络设备运行机制和工作原理加以程序固化，如交换机多生成树收敛过程、VRRP选举机制等。学生对程序看不见、摸不着、想不透，即使学会设备配置方法，也是只知其然而不知其所以然。在实践教学实施过程中将晦涩的理论以生动形象的多媒体动画展现出来，仿真网络运行机制，阐述设备工作原理，将复杂拓扑简单化，抽象理论具体化，配合实物讲解和演示教学，边讲边练，边练边讲，再现配置过程，验证实践结果，既丰富了课程内容，又加深了对知识难点的理解，从而提高了认知水平。

（3）结合“分组实施、制订方案、实践论证”三步实践法开发学生创新思维，提高实践动手能力。

课程基于书中具体工作任务实施分组教学，鼓励学生通过小组讨论、分工合作、角色交替方式共同制订解决方案，论证网络规划，最后通过网络组建和设备配置检验方案的可行性，得出实践结论。让学生知道在解决实际工程问题时，答案没有最好的，方案不是唯一的，以此扩展学生思维，培养学生分析问题和解决问题的能力。

通过对传统教学模式和教学方法的革新，并基于华为eNSP仿真设备的虚拟实践教学改革，既节约了实验室建设投入成本，又提高了学生自主学习兴趣和主动性，解决了传统物理设备配置实验中平时闲置忙时争用的问题。学生完成工作任务时长减少40%以上，教师可以多讲15%的知识点。

由于编著者水平有限，书中难免存在不足之处，恳请广大教师和读者批评、指正。

编著者

2023年3月

目 录

CONTENTS

工作任务一

MSTP 多生成树协议

【工作目的】

理解 MSTP 多生成树协议的工作原理，掌握配置 MSTP 实现负载均衡过程。

【工作任务】

公司在 A 栋楼和 B 栋楼均设置技术部 Vlan10 和工程部 Vlan20，通过二层交换机 SW1 和 SW2 接入。为提高可靠性，并考虑日后服务器网段的接入，新增三层交换机 SW3 和 SW4 以提供冗余链路。提供冗余必然会产生闭合回路，使用传统 STP 协议虽能防止环路产生，但会导致链路闲置问题。为提高链路利用率，公司决定配置 MSTP，在配备冗余的同时达到流量均衡的目的，具体任务如下。

(1) 在 SW1，Vlan10 流量走 SW1-SW3-SW2 之间链路，Vlan20 走 SW1-SW4-SW2 之间链路。

(2) 在 SW2，Vlan10 流量走 SW2-SW3-SW1 之间链路，Vlan20 走 SW2-SW4-SW1 之间链路。

(3) 验证 MSTP 协议可靠性和主备链路切换过程，当备份链路故障时，自动切换到主链路。

【工作背景】

传统生成树协议（spanning-tree）的作用是在交换网络中提供冗余备份链路，并且解决交换网络中环路问题。但由于局域网中所有 Vlan 共享一个生成树，备份链路不承载任何流量，会导致带宽浪费。如图 1-1 所示，在交换机 SW1 和 SW2 之间的主链路承载了所有流量，而备份链路带宽造成了浪费。

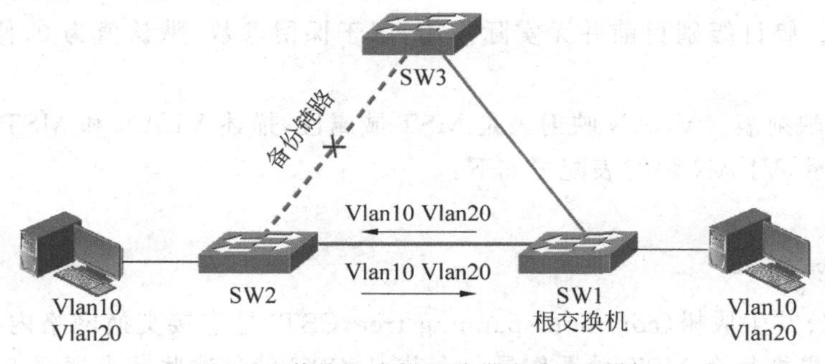

图 1-1　传统生成树链路带宽浪费问题

为解决这一问题，IEEE 于 2002 年发布 802.1S 标准并定义了 MSTP(multiple spanning tree protocol，多生成树协议)。MSTP 把一个交换网络划分成多个域，每个域内形成多棵生成树，生成树之间彼此独立。每棵生成树叫作一个多生成树实例 MSTI(multiple spanning tree instance)，每个域叫作一个 MST 域(multiple spanning tree region)。如图 1-2 所示，生成树实例 1 包含 Vlan10，实例 2 包含 Vlan20，SW1 和 SW3 分别作为实例 1 和实例 2 的根交换机。对于实例 1，链路 2 是备份链路，Vlan10 流量经链路 1 转发；对于实例 2，链路 1 是备份链路，Vlan20 流量经链路 2 和链路 3 转发；当链路 2 或链路 3 故障时，所有 Vlan 流量都经链路 1 转发，从而达到冗余备份和负载均衡的目的。

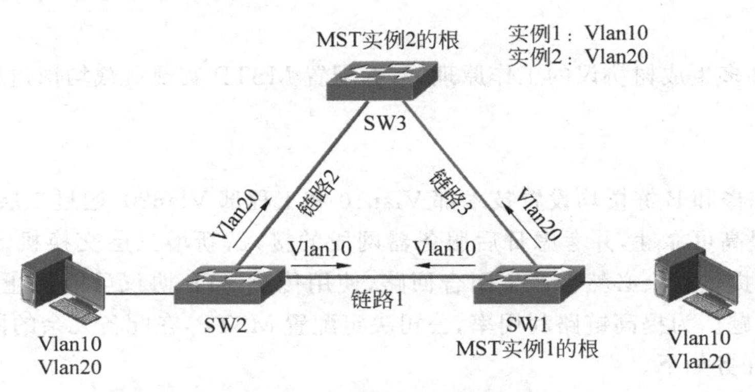

图 1-2 MSTP 流量均衡解决方案

【任务分析】

多生成树专业术语如下。

(1) MST 域(MST region)。一个局域网可以存在多个 MST 域，各 MST 域之间在物理上直接或间接相连。用户可以通过 MSTP 配置命令把多台交换设备划分在同一个 MST 域内。以下为交换机处于相同 MST 域的条件。

① 所有交换机都启动 MSTP 模式(华为交换机 STP 默认 MSTP 模式)(stp mode mstp)。

② 具有相同域名(region-name STRING＜1-32＞)。

③ 具有相同 Vlan 实例映射表。

④ 具有相同 MSTP 修订级别配置(revision-level INTEGER＜0-65535＞)。

注意：MSTP 的修订级别用来与 MST 域名和 MST 域的 VLAN 映射表共同确定设备所属的 MST 域。修订级别目前并无实际作用，属于保留参数，默认值为 0，修订级别属于可选项。

(2) VLAN 映射表。VLAN 映射表是 MST 域属性，描述 VLAN 和 MSTI 之间的映射关系。如图 1-2 所示，VLAN 映射表配置如下：

```
instance 1 vlan 10
instance 2 vlan 20
```

(3) CST。公共生成树(common spanning tree，CST)是连接交换网络内所有 MST 域的一棵生成树。如果把每个 MST 域看作是一个节点，CST 就是这些节点通过 STP 或 RSTP 协议计算生成的一棵生成树。

(4) CIST。公共和内部生成树(common and internal spanning tree，CIST)是通过 STP

或 RSTP 协议计算生成的,连接一个交换网络内所有交换设备的单生成树。

(5) 域根。域根(regional root)分为 IST 域根和 MSTI 域根。一个 MST 域内可以生成多棵生成树,每棵生成树都称为一个 MSTI。MSTI 域根是每个多生成树实例的树根。

(6) 总根。总根是 CIST(common and internal spanning tree)的根桥。

(7) 端口角色。端口角色见表 1-1。

表 1-1 端口角色

端口角色	说明
根端口	在非根桥上,离根桥最近的端口是本交换设备的根端口。根交换设备没有根端口。根端口负责向树根方向转发数据
指定端口	对一台交换设备而言,指定端口是向下游交换设备转发 BPDU 报文的端口
Alternate 端口	从配置 BPDU 报文发送角度来看,Alternate 端口是由于学习到其他网桥发送的配置 BPDU 报文而阻塞的端口。从用户流量角度来看,Alternate 端口提供从指定桥到根的另一条可切换路径,作为根端口的备份端口
Backup 端口	从配置 BPDU 报文发送角度来看,Backup 端口是由于学习到自己发送的配置 BPDU 报文而阻塞的端口。从用户流量角度来看,Backup 端口作为指定端口的备份,提供另外一条从根节点到叶节点的备份通路
Master 端口	Master 端口是 MST 域和总根相连的所有路径中最短路径上的端口,它是交换设备上连接 MST 域到总根的端口;Master 端口是域中的报文去往总根的必经之路;Master 端口是特殊域边缘端口,Master 端口在 CIST 上的角色是 Root Port,在其他各实例上的角色都是 Master 端口
域边缘端口	域边缘端口是指位于 MST 域的边缘并连接其他 MST 域或 SST 的端口
边缘端口	如果指定端口位于整个域的边缘,不再与任何交换设备连接,这种端口叫作边缘端口;边缘端口一般与用户终端设备直接连接,不参与生成树选举过程

(8) 端口状态。MSTP 定义的端口状态与 RSTP 协议中定义相同,共有 disable、blocking、listening、learning、forwarding 5 个状态。

【设备器材】

接入层交换机(S3700)2 台,汇聚层交换机(S5700)2 台。

主机 4 台,承担角色见表 1-2。

表 1-2 主机配置表

角色	接入方式	IP 地址	Vlan ID
主机 1	eNSP PC 接入	192.168.10.10/24	Vlan10(技术部)
主机 2	eNSP PC 接入	192.168.20.10/24	Vlan20(工程部)
主机 3	eNSP PC 接入	192.168.10.20/24	Vlan10(技术部)
主机 4	eNSP PC 接入	192.168.20.20/24	Vlan20(工程部)

【环境拓扑】

环境拓扑如图 1-3 所示。

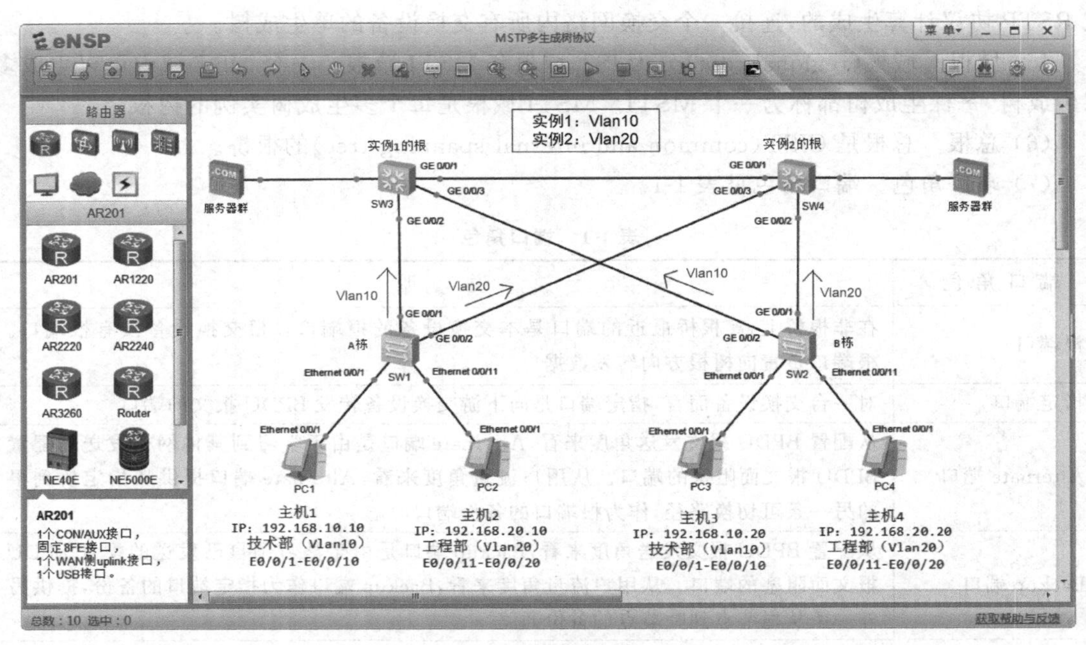

图 1-3 环境拓扑

【工作过程】

1. 基本配置

(1) 交换机 Vlan 和端口配置。

```
<Huawei>                                                         //用户视图
<Huawei>system-view                                              //进入系统视图
[Huawei]sysname SW1                                              //更改设备名称
[SW1]vlan batch 10 20                                            //batch:批量
[SW1]port-group 1                                                //技术部组
[SW1-port-group-1]group-member Ethernet 0/0/1 to Ethernet 0/0/10
[SW1-port-group-1]port link-type access
[SW1-port-group-1]port default vlan 10
[SW1-port-group-1]quit
[SW1]port-group 2                                                //工程部组
[SW1-port-group-2]group-member Ethernet 0/0/11 to Ethernet 0/0/20
[SW1-port-group-2]port link-type access
[SW1-port-group-2]port default vlan 20
[SW1-port-group-2]quit
[SW1]port-group 3                                                //Trunk 组
[SW1-port-group-3]group-member GigabitEthernet 0/0/1 GigabitEthernet 0/0/2
[SW1-port-group-3]port link-type trunk
[SW1-port-group-3]port trunk allow-pass vlan 10 20
                          //trunk 口本身应允许所有 Vlan 通过，一般写为
                            "port trunk allow-pass vlan all"，也可以
                            手动指定允许哪些 Vlan 通过
[SW1-port-group-3]quit
```

```
[SW1]
-----------------------------------------------------------------------------------
<Huawei>system-view
[Huawei]sysname SW2
[SW2]vlan batch 10 20
[SW2]port-group 1
[SW2-port-group-1]group-member Ethernet 0/0/1 to Ethernet 0/0/10   //技术部组
[SW2-port-group-1]port link-type access
[SW2-port-group-1]port default vlan 10
[SW2-port-group-1]quit
[SW2]port-group 2
[SW2-port-group-2]group-member Ethernet 0/0/11 to Ethernet 0/0/20  //工程部组
[SW2-port-group-2]port link-type access
[SW2-port-group-2]port default vlan 20
[SW2-port-group-2]quit
[SW2]port-group 3                                                  //Trunk 组
[SW2-port-group-3]group-member GigabitEthernet 0/0/1 GigabitEthernet 0/0/2
[SW2-port-group-3]port link-type trunk
[SW2-port-group-3]port trunk allow-pass vlan all
[SW2-port-group-3]quit
[SW2]
-----------------------------------------------------------------------------------
<Huawei>system-view
[Huawei]sysname SW3
[SW3]vlan batch 10 20
[SW3]port-group 1                                                  //Trunk 组
[SW3-port-group-1]group-member GigabitEthernet 0/0/1 to GigabitEthernet 0/0/3
[SW3-port-group-1]port link-type trunk
[SW3-port-group-1]port trunk allow-pass vlan all
[SW3-port-group-1]quit
[SW3]
-----------------------------------------------------------------------------------
<Huawei>system-view
[SW3]sysname SW4
[SW4]vlan batch 10 20
[SW4]port-group 1
[SW4-port-group-1]group-member GigabitEthernet 0/0/1 to GigabitEthernet 0/0/3
[SW4-port-group-1]port link-type trunk
[SW4-port-group-1]port trunk allow-pass vlan all
[SW4-port-group-1]quit
[SW4]
```

(2) 创建 MSTP 域,配置实例 Vlan 映射关系。

```
[SW1]stp mode mstp                          //华为交换机 STP 默认 MSTP 模式,本行可以不输入
[SW1]stp region-configuration               //进入 MSTP 域视图
[SW1-mst-region]region-name gdcp_company    //自定义域名
[SW1-mst-region]revision-level 1            //指定 MSTP 修订级别为 1,如用默认 0 本行可不输入
[SW1-mst-region]instance 1 vlan 10          //MSTP 实例 1 包含 vlan 10
[SW1-mst-region]instance 2 vlan 20          //MSTP 实例 1 包含 vlan 20
```

```
[SW1-mst-region]active region-configuration   //激活域配置
Info: This operation may take a few seconds. Please wait for a moment...done.
[SW1-mst-region]quit
[SW1]
```

注意：对其他交换机做同样配置，在同一域中，必须有相同的域名 gdcp_company、修订级别 1 和实例 Vlan 映射关系。

```
[SW2]stp mode mstp
[SW2]stp region-configuration
[SW2-mst-region]region-name gdcp_company
[SW2-mst-region]revision-level 1
[SW2-mst-region]instance 1 vlan 10
[SW2-mst-region]instance 2 vlan 20
[SW2-mst-region]active region-configuration
Info: This operation may take a few seconds. Please wait for a moment...done.
[SW2-mst-region]quit
[SW2]
```

```
[SW3]stp mode mstp
[SW3]stp region-configuration
[SW3-mst-region]region-name gdcp_company
[SW3-mst-region]revision-level 1
[SW3-mst-region]instance 1 vlan 10
[SW3-mst-region]instance 2 vlan 20
[SW3-mst-region]active region-configuration
Info: This operation may take a few seconds. Please wait for a moment...done.
[SW3-mst-region]quit
[SW3]
```

```
[SW4]stp mode mstp
[SW4]stp region-configuration
[SW4-mst-region]region-name gdcp_company
[SW4-mst-region]revision-level 1
[SW4-mst-region]instance 1 vlan 10
[SW4-mst-region]instance 2 vlan 20
[SW4-mst-region]active region-configuration
Info: This operation may take a few seconds. Please wait for a moment...done.
[SW4-mst-region]quit
[SW4]
```

2. 任务验证

(1) 未指定 MST 根交换机时，连通性验证。

在主机 1 命令行中输入 ping 192.168.10.20 -t 反复测试主机 3 连通性，在主机 2 命令行中输入 ping 192.168.20.20 -t 反复测试主机 4 连通性，结果如图 1-4 所示。

(2) 未指定 MST 根交换机时，验证转发路径。

保持如图 1-4 所示连通性测试状态，分别在交换机 SW1 的 G0/0/1 和 G0/0/2 接口启用 Wireshark 抓包程序，如图 1-5 所示。可以看到主机 1 和主机 2 流量经 SW1 的 G0/0/1 接口转

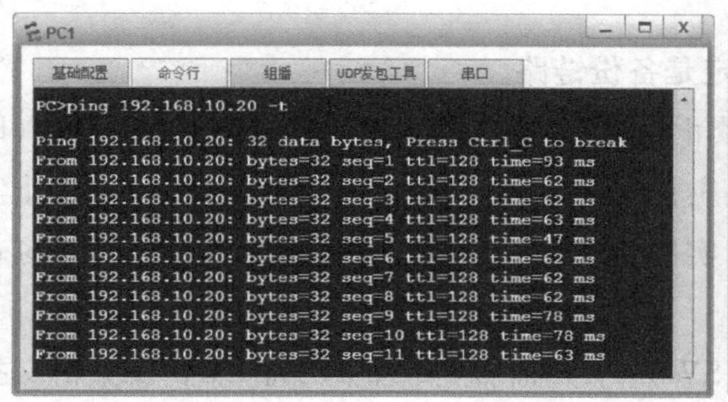

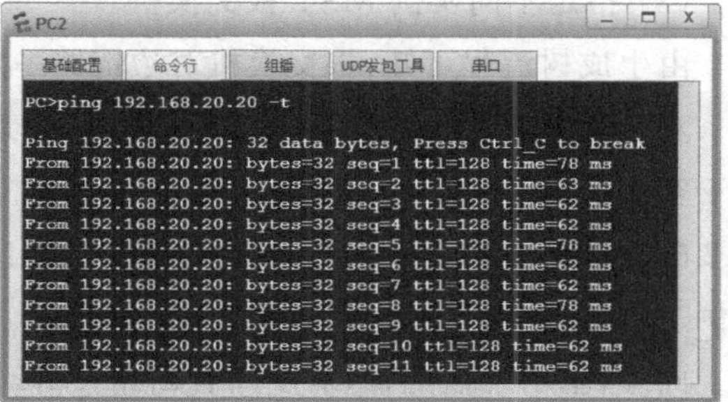

图 1-4　测试主机 1、主机 2 到主机 3、主机 4 的连通性

发，G0/0/2 接口接收到上行接口周期发送的 BPDU 包，并不转发任何流量，处于闲置状态，如图 1-6 所示。虽然配置了 MSTP 多生成树协议，每个实例独立计算生成树，但是在总拓扑结构和参数相同情况下，不同实例生成树选举结果是一致的，导致所有 Vlan 流量挤在同一通路。

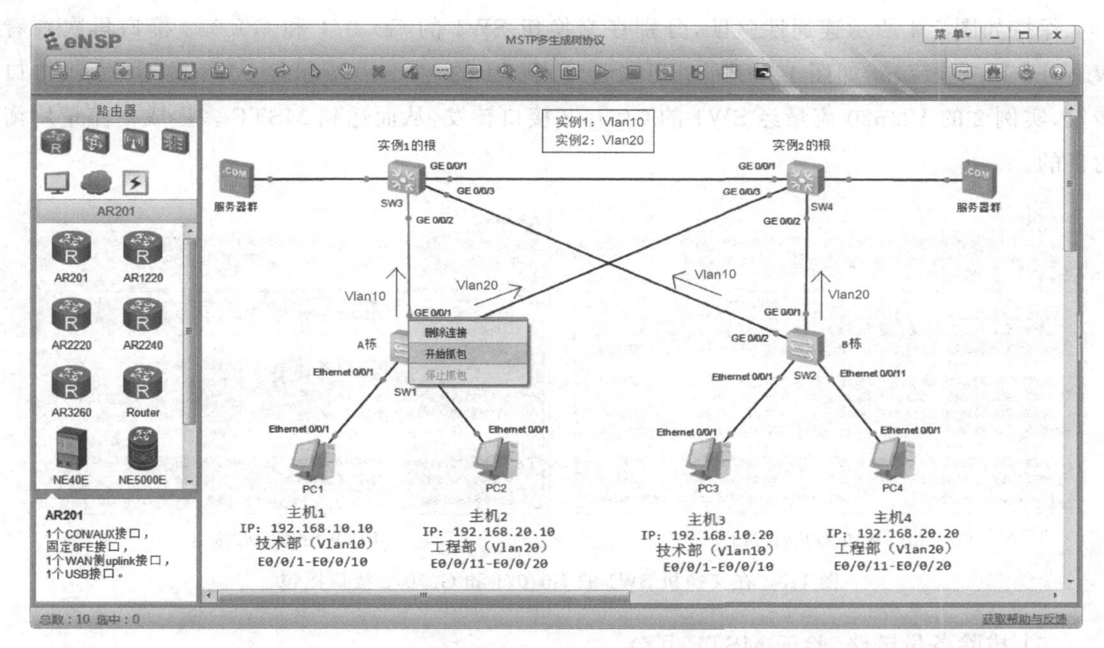

图 1-5　在交换机 SW1 的 G0/0/1 和 G0/0/2 接口分别启用抓包程序

No.	Time	Source	Destination	Protocol
969	229.867000	192.168.10.20	192.168.10.10	ICMP
970	229.898000	192.168.20.20	192.168.20.10	ICMP
971	230.897000	192.168.10.10	192.168.10.20	ICMP
972	230.912000	192.168.20.10	192.168.20.20	ICMP
973	230.944000	192.168.10.20	192.168.10.10	ICMP
974	230.959000	192.168.20.20	192.168.20.10	ICMP
975	231.084000	HuaweiTe_74:1a:01	Spanning-tree-(for-…	STP
976	231.973000	192.168.10.10	192.168.10.20	ICMP
977	231.989000	192.168.20.10	192.168.20.20	ICMP

(a) SW1的G0/0/1接口

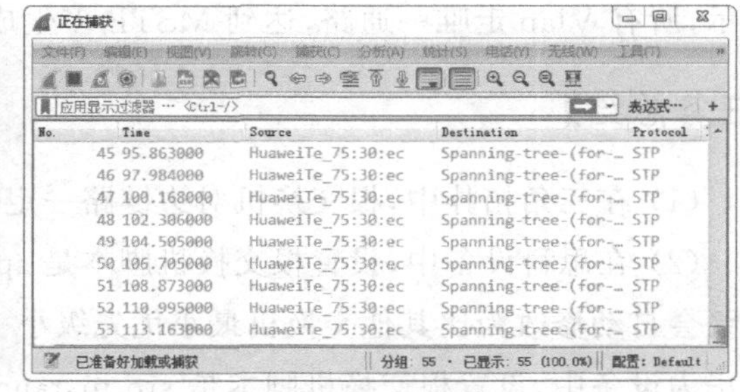

No.	Time	Source	Destination	Protocol
45	95.863000	HuaweiTe_75:30:ec	Spanning-tree-(for-…	STP
46	97.984000	HuaweiTe_75:30:ec	Spanning-tree-(for-…	STP
47	100.168000	HuaweiTe_75:30:ec	Spanning-tree-(for-…	STP
48	102.306000	HuaweiTe_75:30:ec	Spanning-tree-(for-…	STP
49	104.505000	HuaweiTe_75:30:ec	Spanning-tree-(for-…	STP
50	106.705000	HuaweiTe_75:30:ec	Spanning-tree-(for-…	STP
51	108.873000	HuaweiTe_75:30:ec	Spanning-tree-(for-…	STP
52	110.995000	HuaweiTe_75:30:ec	Spanning-tree-(for-…	STP
53	113.163000	HuaweiTe_75:30:ec	Spanning-tree-(for-…	STP

(b) SW1的G0/0/2接口

图 1-6　在交换机 SW1 的 G0/0/1 和 G0/0/2 接口抓包

(3) 指定不同实例 MST 根交换机。

由生成树流程可知,根交换机对边链路一定是备份链路。

对于闭合回路(SW1-SW3-SW4)而言,为禁止实例 1 的 Vlan10 走 SW1-SW4 之间链路(即对于实例 1,SW1-SW4 之间链路为备份链路),应在实例 1 中将 SW3 设为根交换机,为禁止实例 2 的 Vlan20 走 SW1-SW3 之间链路(即对于实例 2,SW1-SW3 之间链路为备份链路),应在实例 2 中将 SW4 设为根交换机。

同理,基于上述配置,对于闭合回路(SW2-SW3-SW4)而言,实例 1 已将 SW3 设为根交换机,则禁止 Vlan10 走 SW2-SW4 之间链路,实例 2 已将 SW4 设为根交换机,则禁止 Vlan20 走 SW2-SW3 之间链路,从而实现流量平衡,以下为配置脚本。

```
[SW3]stp instance 1 root primary          //等价于:stp instance 1 priority 0
[SW4]stp instance 2 root primary          //等价于:stp instance 2 priority 0
```

(4) 指定 MST 根交换机后,验证 MSTP 流量均衡。

保持如图 1-4 所示连通性验证,分别在交换机 SW1 的 G0/0/1 和 G0/0/2 接口再次查看 Wireshark 抓包程序,如图 1-7 所示。可以看到实例 1 的 Vlan10 流量经 SW1 的 G0/0/1 接口转发,实例 2 的 Vlan20 流量经 SW1 的 G0/0/2 接口转发,从而达到 MSTP 多生成树流量均衡的目的。

No.	Time	Source	Destination	Protocol
9556	2346.785000	192.168.10.20	192.168.10.10	ICMP
9557	2347.815000	192.168.10.10	192.168.10.20	ICMP
9558	2347.862000	192.168.10.20	192.168.10.10	ICMP
9559	2348.142000	HuaweiTe_74:1a:01	Spanning-tree-(for-…	STP
9560	2348.876000	192.168.10.10	192.168.10.20	ICMP
9561	2348.938000	192.168.10.20	192.168.10.10	ICMP
9562	2349.936000	192.168.10.10	192.168.10.20	ICMP
9563	2349.999000	192.168.10.20	192.168.10.10	ICMP
9564	2350.295000	HuaweiTe_74:1a:01	Spanning-tree-(for-…	STP
9565	2351.013000	192.168.10.10	192.168.10.20	ICMP
9566	2351.060000	192.168.10.20	192.168.10.10	ICMP

Frame (frame), 151 字节 分组: 9566 · 已显示: 9566 (100.0%) 配置: Default

(a) SW1的G0/0/1接口

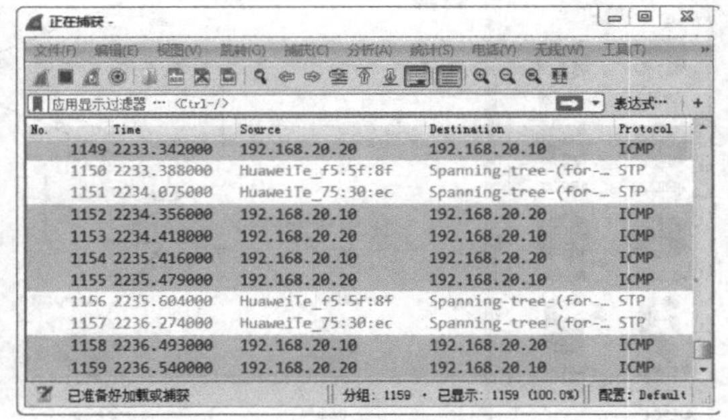

No.	Time	Source	Destination	Protocol
1149	2233.342000	192.168.20.20	192.168.20.10	ICMP
1150	2233.388000	HuaweiTe_f5:5f:8f	Spanning-tree-(for-…	STP
1151	2234.075000	HuaweiTe_75:30:ec	Spanning-tree-(for-…	STP
1152	2234.356000	192.168.20.10	192.168.20.20	ICMP
1153	2234.418000	192.168.20.20	192.168.20.10	ICMP
1154	2235.416000	192.168.20.10	192.168.20.20	ICMP
1155	2235.479000	192.168.20.20	192.168.20.10	ICMP
1156	2235.604000	HuaweiTe_f5:5f:8f	Spanning-tree-(for-…	STP
1157	2236.274000	HuaweiTe_75:30:ec	Spanning-tree-(for-…	STP
1158	2236.493000	192.168.20.10	192.168.20.20	ICMP
1159	2236.540000	192.168.20.20	192.168.20.10	ICMP

已准备好加载或捕获 分组: 1159 · 已显示: 1159 (100.0%) 配置: Default

(b) SW1的G0/0/2接口

图 1-7 在交换机 SW1 的 G0/0/1 和 G0/0/2 接口抓包

(5) 拔除备份链路,验证 MSTP 冗余。

拔除实例 1 和实例 2 备份链路,拓扑结构如图 1-8 所示,不影响主机之间连通性,如图 1-9 所示,所有 Vlan 走唯一通路,达到 MSTP 多生成树冗余备份的目的。

【任务总结】

(1) 在三角拓扑中,根交换机对边链路一定作为备份链路。

(2) 在思科设备中,设置根交换机脚本是 spanning-tree vlan 10 root primary,根交换机优先级会自动修改为比其他交换机最小优先级小一点的值,值为 4096 倍数,但具体值无法预测;在华为设备中,设置根交换机脚本是 stp instance 1 root primary(具体 Vlan 值须在实例中事先指定),根交换机优先级将设置为 0。

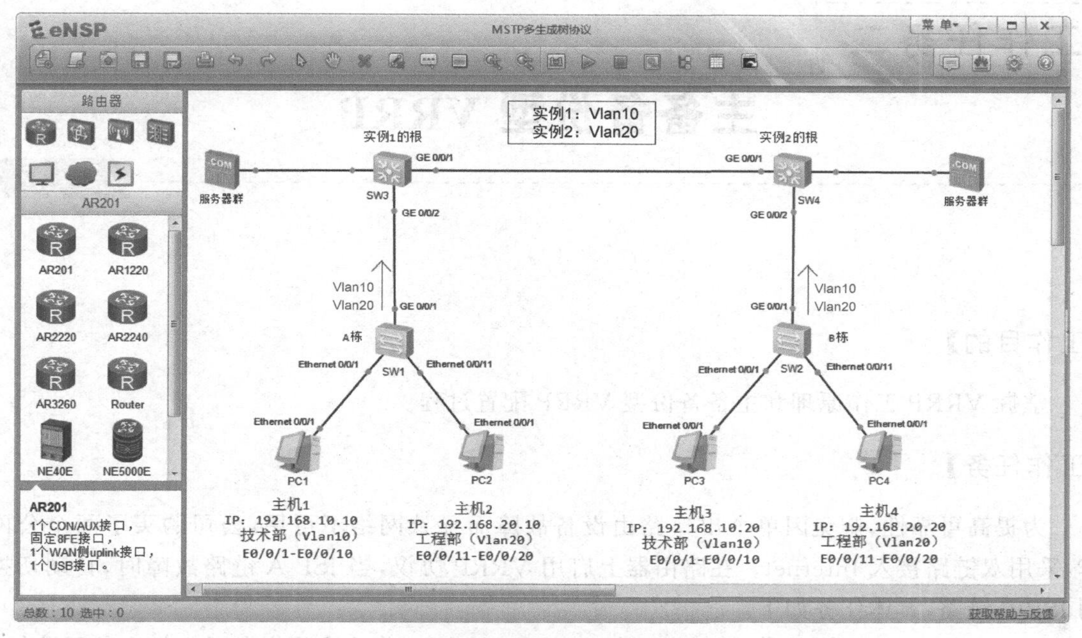

图 1-8 拔除实例 1 和实例 2 备份链路

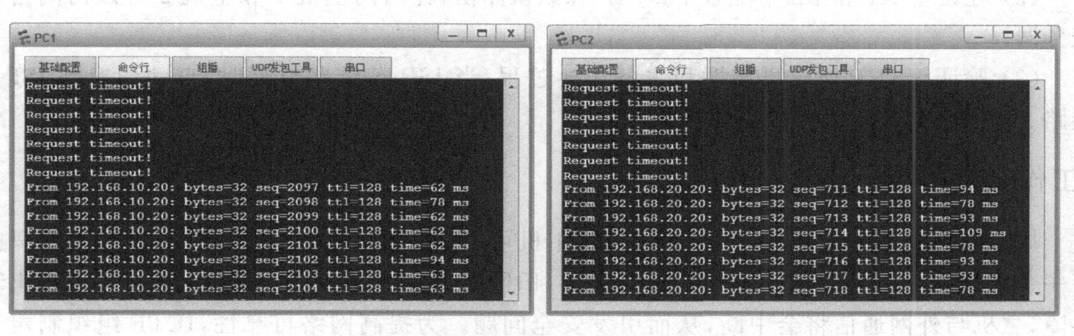

图 1-9 拔除备份链路不影响连通性

工作任务二

主备备份型 VRRP

【工作目的】

掌握 VRRP 工作原理和主备备份型 VRRP 配置过程。

【工作任务】

为提高可靠性，避免因单个出口路由设备故障而导致网络不可用，公司购买了两个公网 IP，采用双链路接入 Internet。在路由器上启用 VRRP 协议，当 ISP-A 链路故障时，自动切换到 ISP-B 链路，具体任务如下。

(1) 建立 VRRP 分组，将 R1 与 R2 加入 VRRP 组，R1 作为主用设备，R2 作为备用设备。

(2) 在设备 R1 和 R2 中配置 Easy-IP 和默认路由，让内网主机 1 和主机 2 可以访问公网服务器。

(3) 验证 VRRP 协议可靠性和主备切换过程，当 ISP-A 链路故障时，自动切换到 ISP-B 链路。

【工作背景】

局域网内网主机只能配置一个网关 IP（主机可以配置多个不同网段 IP，但同一时刻只能有一个默认网关 IP，以新配置的网关 IP 为准，旧网关 IP 自动失效）。假如出口网关设备出现故障，主机与外网通信将会中断，从而引发安全问题。为提高网络可靠性，IETF 组织制定了 VRRP（虚拟路由冗余协议），利用多个路由器组建虚拟网关设备，避免因单个路由器故障而导致外部网络的不可用。

【任务分析】

VRRP（virtual router redundancy protocol，虚拟路由冗余协议）是一种容错协议，用于解决局域网内部主机访问外部公网的可靠性问题。VRRP 将多个路由器逻辑上组成一个虚拟出口网关设备，该虚拟设备（由路由器组成）可以配置新的 IP 地址或 Router ID。路由器之间通过选举产生一个 Master 路由器（主用设备）和多个 Backup 路由器（备用设备），主用设备负责转发业务流量，备用设备负责监听主用设备状态。当主用设备发生故障时，备用设备（假如有多个备用设备，仍需要选举产生一个新的主用设备）自动承接主用设备工作，转发业务流量，从而保证内外网通信的持续性和可靠性。

VRRP 通告报文使用 IP 多播数据包进行封装，组地址为 224.0.0.18，路由器通告彼此优先级以选举主用设备和备用设备。VRRP 协议优先级范围＜0～255＞，可以配置的范围是＜1～254＞，默认值 100，数值越大优先级越高。

【设备器材】

汇聚层交换机(S5700)1 台，路由器(AR1220)3 台，需添加 1GEC 接口。

主机 3 台，承担角色见表 2-1。

表 2-1　主机配置表

角　　色	接 入 方 式	IP 地址	网　　关
主机 1	eNSP PC 接入	192.168.1.10/24	192.168.1.1
主机 2	eNSP PC 接入	192.168.1.20/24	192.168.1.1
公网服务器	eNSP Server 接入	112.32.32.200/24	112.32.32.1

【环境拓扑】

环境拓扑如图 2-1 所示。

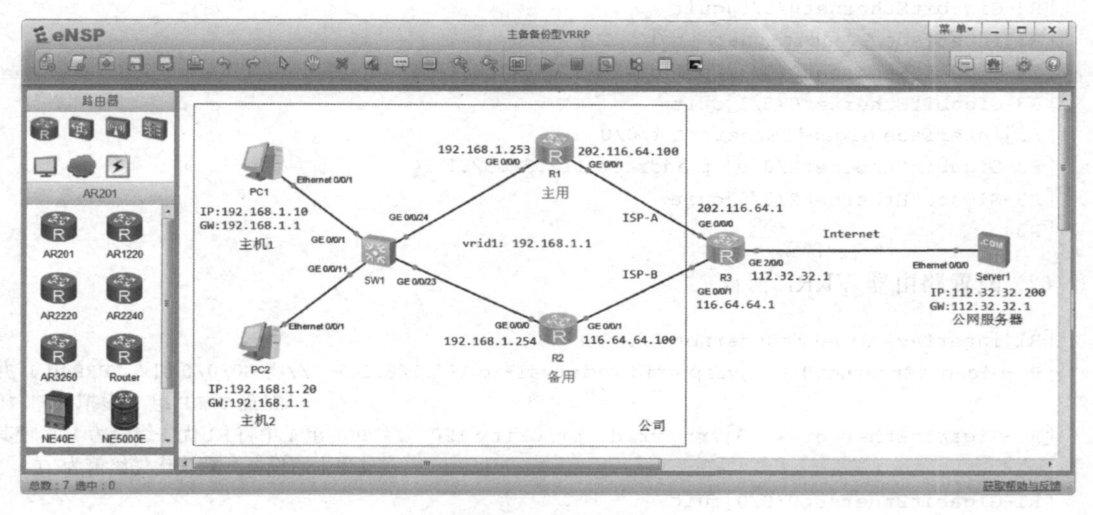

图 2-1　环境拓扑

【工作过程】

1. 基本配置

(1) 路由器接口 IP 配置。

```
<Huawei>                                               //用户视图
<Huawei>system-view                                    //进入系统视图
[Huawei]sysname R1
[R1]interface GigabitEthernet 0/0/0
[R1-GigabitEthernet0/0/0]ip address 192.168.1.253 255.255.255.0
[R1-GigabitEthernet0/0/0]quit
[R1]interface GigabitEthernet 0/0/1
[R1-GigabitEthernet0/0/1]ip address 202.116.64.100 255.255.255.0
[R1-GigabitEthernet0/0/1]quit
[R1]
```

```
<Huawei>
<Huawei>system-view
[Huawei]sysname R2
[R2]interface GigabitEthernet 0/0/0
[R2-GigabitEthernet0/0/0]ip address 192.168.1.254 24          //24 表示子网掩码
[R2-GigabitEthernet0/0/0]interface GigabitEthernet 0/0/1      //可直接进入其他接口,但
                                                                脚本不能自动补齐
[R2-GigabitEthernet0/0/1]ip address 116.64.64.100 24
[R2-GigabitEthernet0/0/1]quit
[R2]

<Huawei>
<Huawei>system-view
[Huawei]sysname R3
[R3]interface GigabitEthernet 0/0/0
[R3-GigabitEthernet0/0/0]ip address 202.116.64.1 24
[R3-GigabitEthernet0/0/0]quit
[R3]interface GigabitEthernet 0/0/1
[R3-GigabitEthernet0/0/1]ip address 116.64.64.1 24
[R3-GigabitEthernet0/0/1]quit
[R3]interface GigabitEthernet 2/0/0
[R3-GigabitEthernet2/0/0]ip address 112.32.32.1 24
[R3-GigabitEthernet2/0/0]quit
[R3]
```

(2) 配置路由器 VRRP 协议。

```
[R1]interface GigabitEthernet 0/0/0
[R1-GigabitEthernet0/0/0]vrrp vrid 1 virtual-ip 192.168.1.1   //将 G0/0/0 加入 VRRP 组 1,并
                                                                配置 VRRP 组 1 虚拟接口 IP
[R1-GigabitEthernet0/0/0]vrrp vrid 1 priority 120   //VRRP 组 1 中将 R1 优先级设为 120,使其
                                                      成为主用设备。默认优先级值为 100
[R1-GigabitEthernet0/0/0]quit
[R1]
--------------------------------------------------------------------------------
[R2]interface GigabitEthernet 0/0/0
[R2-GigabitEthernet0/0/0]vrrp vrid 1 virtual-ip 192.168.1.1
[R2-GigabitEthernet0/0/0]quit
[R2]
```

注意:

① 优先级大小决定路由器主用设备或备用设备角色。优先级高(优先级大)的选举为主用设备 Master,反之优先级低(优先级小)的选举为备用设备 Backup。如果优先级相同,则接口 IP 大的选举为 Master。

② R1 和 R2 必须处于同一个 VRRP 组中,如本例的 vrid 1。

③ 在路由器中,VRRP 虚拟接口 IP 必须和当前接口 IP 处于同一网段,如本例即都在<192.168.1.0>网段。假如将 vrid 1 的 IP 配置为"10.1.1.1",系统会提示以下错误。

```
[R1-GigabitEthernet0/0/0]vrrp vrid 1 virtual-ip 10.1.1.1
Error: The virtual IP address is not within a subnet on this interface.
```

(3) 在 R1 主用设备配置上行接口监视。

```
[R1]interface GigabitEthernet 0/0/0
[R1 - GigabitEthernet0/0/0] vrrp vrid 1 track interface GigabitEthernet 0/0/1
reduced 30                                  //在 VRRP 组 1 中监视上行接口 G0/0/1,假
                                              如该端口 Down,则将当前优先级减 30,即
                                              120-30=90(优先级小于 R2 的默认值 100),
                                              让 R2 成为主用设备
```

注意:

① VRRP 可以监视本地接口,也可以监视一个 IP(一般为非本地接口 IP)。当指定 IP 不可达时触发优先级变化,例如:

```
[R1-GigabitEthernet0/0/0]vrrp vrid 1 track ip route 202.116.64.1 24 reduced 30
```

② reduced 参数用于减少优先值,increased 参数用于增加优先值。

(4) 配置路由器 Easy-IP 和默认路由。

```
[R1]acl 2000          //基本 ACL:<2000～2999>,只能根据源 IP 地址过滤
                      //高级 ACL:<3000～3999>,基于源 IP、目的 IP、协议类型等过滤,类似扩展 ACL
[R1-acl-basic-2000]rule permit source 192.168.1.0 0.0.0.255
[R1-acl-basic-2000]quit
[R1]interface GigabitEthernet 0/0/1
[R1-GigabitEthernet0/0/1]nat outbound 2000   //加载 ACL2000 过滤规则与公网接口出栈之间
                                               的转换关系,即把内网 IP 经过滤规则匹配后
                                               转换为公网接口 IP
[R1-GigabitEthernet0/0/1]quit
[R1]ip route-static 0.0.0.0 0.0.0.0 202.116.64.1
[R1]
---------------------------------------------------------------------------------
[R2]acl 2000
[R2-acl-basic-2000]rule permit source any
[R2-acl-basic-2000]quit
[R2]interface GigabitEthernet 0/0/1
[R2-GigabitEthernet0/0/1]nat outbound 2000
[R2-GigabitEthernet0/0/1]quit
[R2]ip route-static 0.0.0.0 0.0.0.0 116.64.64.1
[R2]
```

2. 任务验证

(1) VRRP 主备备份模式验证。

在路由器上查看 VRRP 组详细状态。

```
[R1]display vrrp
  GigabitEthernet0/0/0 | Virtual Router 1
    State : Master                          //R1 选举为主用设备
    Virtual IP : 192.168.1.1                //VRRP 组 1 的 IP
    Master IP : 192.168.1.253               //本地 G0/0/0 接口 IP
    PriorityRun : 120                       //当前优先级值
    PriorityConfig : 120
    MasterPriority : 120                    //VRRP 组 1 中主用设备的优先级值
    Preempt : YES   Delay Time : 0 s
    TimerRun : 1 s
```

```
    TimerConfig : 1 s
    Auth type : NONE
    Virtual MAC : 0000-5e00-0101          //VRRP 组 1 虚拟接口 MAC 地址
    Check TTL : YES
    Config type : normal-vrrp
    Backup-forward : disabled
    Create time : 2021-07-19 17:01:18 UTC-08:00
    Last change time : 2021-07-19 17:34:39 UTC-08:00
-----------------------------------------------------------------------------------
[R2]display vrrp
  GigabitEthernet0/0/0 | Virtual Router 1
    State : Backup                        //R2 选举为备用设备
    Virtual IP : 192.168.1.1
    Master IP : 192.168.1.253
    PriorityRun : 100                     //当前优先级值
    PriorityConfig : 100
    MasterPriority : 120                  //VRRP 组 1 中主用设备的优先级值
    Preempt : YES   Delay Time : 0 s
    TimerRun : 1 s
    TimerConfig : 1 s
    Auth type : NONE
    Virtual MAC : 0000-5e00-0101          //与 R1 查询到的 VRRP 组 1 虚拟接口 MAC 地址一致
    Check TTL : YES
    Config type : normal-vrrp
    Backup-forward : disabled
    Create time : 2021-07-19 17:06:19 UTC-08:00
    Last change time : 2021-07-19 17:35:07 UTC-08:00
```

(2) 测试与公网服务器的连通性。

在主机 1 或主机 2 上的命令行窗口 ping 公网服务器，TTL 值为 253，途经 2 个路由器转发，连通性验证如图 2-2 所示。

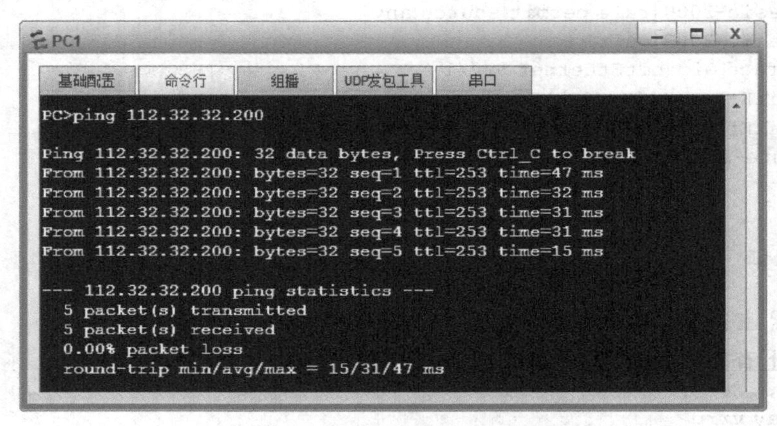

图 2-2 主机 1 可以连通公网服务器

(3) 测试访问公网服务器时数据包转发路径。

在主机 1 或主机 2 的命令行窗口运行 tracert 命令，可以验证与外网通信途经主用设备 R1 转发(走 ISP-A)，如图 2-3 所示。

(4) 验证 VRRP 主备切换。

拔除 R1 与 R3 之间线缆，拓扑图如图 2-4 所示。主机 1 或主机 2 仍可以连通公网服务器，

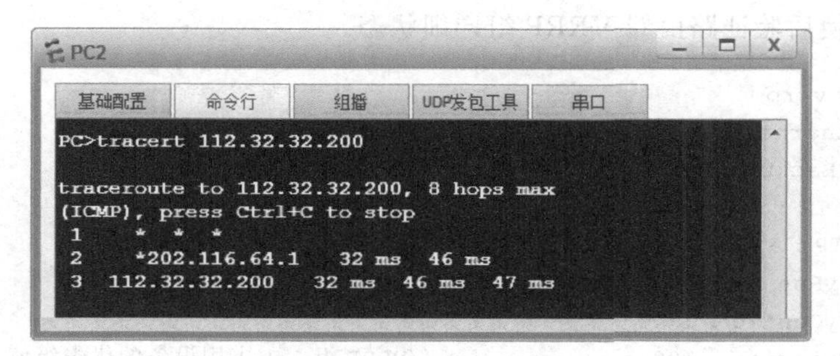

图 2-3　主机 2 数据包经 R1 转发

运行 tracert 命令结果如图 2-5 所示，可以验证所有主机与外网通信途经 R2 转发（走 ISP-B），其中超时是由于路由器重新选举主用设备和备用设备所致。

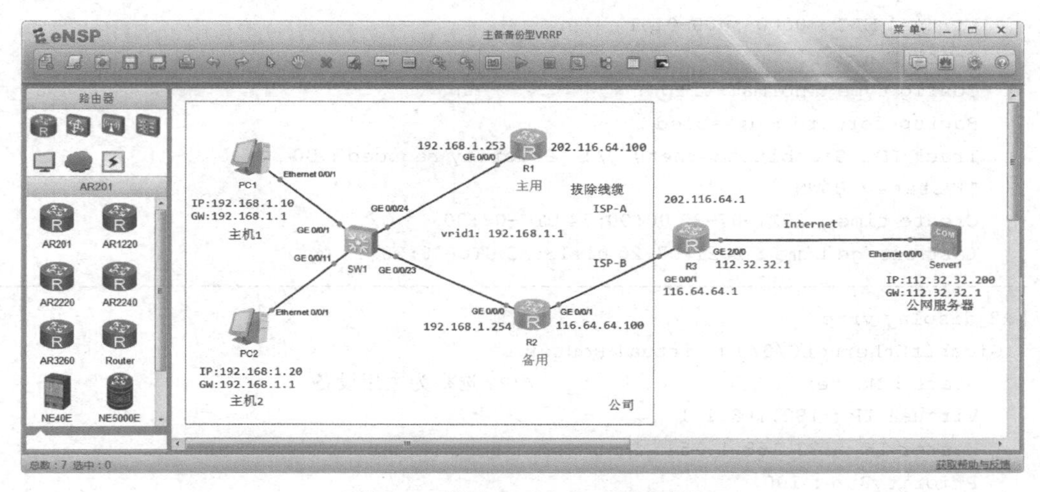

图 2-4　拔除 R1 与 R3 之间线缆

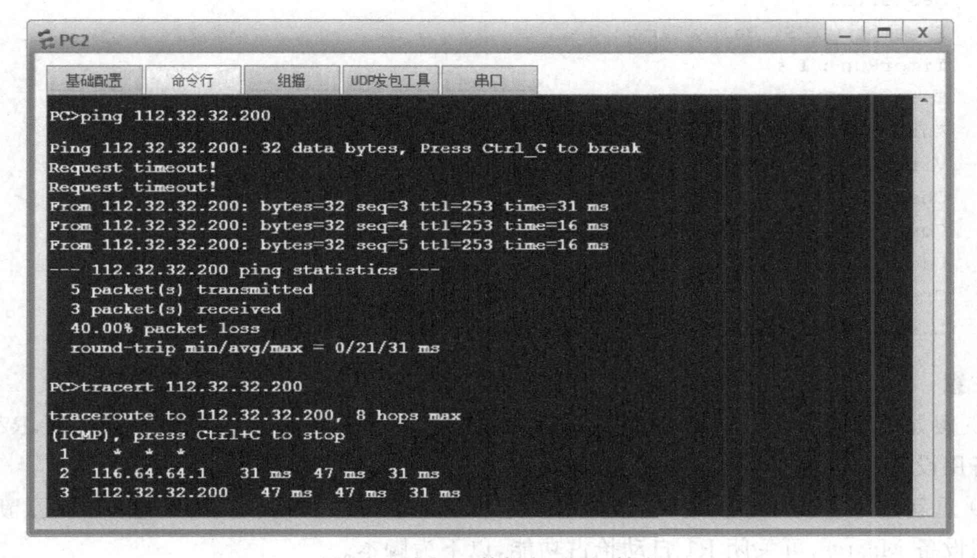

图 2-5　主机 2 数据包经 R2 转发

(5) 主备切换后验证路由器 VRRP 组详细状态。

```
[R1]display vrrp
  GigabitEthernet0/0/0 | Virtual Router 1
    State : Backup                          //R1 选举为备用设备
    Virtual IP : 192.168.1.1
    Master IP : 192.168.1.254
    PriorityRun : 90                        //优先级为 120-30=90
    PriorityConfig : 120
    MasterPriority : 100                    //VRRP 组 1 中主用设备的优先级值
    Preempt : YES   Delay Time : 0 s
    TimerRun : 1 s
    TimerConfig : 1 s
    Auth type : NONE
    Virtual MAC : 0000-5e00-0101
    Check TTL : YES
    Config type : normal-vrrp
    Backup-forward : disabled
    Track IF : GigabitEthernet0/0/1  Priority reduced : 30
    IF state : DOWN
    Create time : 2021-07-20 00:00:44 UTC-08:00
    Last change time : 2021-07-20 01:13:33 UTC-08:00
--------------------------------------------------------------------------------
[R2]display vrrp
  GigabitEthernet0/0/0 | Virtual Router 1
    State : Master                          //R2 选举为主用设备
    Virtual IP : 192.168.1.1
    Master IP : 192.168.1.254
    PriorityRun : 100
    PriorityConfig : 100
    MasterPriority : 100
    Preempt : YES   Delay Time : 0 s
    TimerRun : 1 s
    TimerConfig : 1 s
    Auth type : NONE
    Virtual MAC : 0000-5e00-0101
    Check TTL : YES
    Config type : normal-vrrp
    Backup-forward : disabled
    Create time : 2021-07-20 00:00:47 UTC-08:00
    Last change time : 2021-07-20 01:13:33 UTC-08:00
```

注意：

① 假如重新连接 R1 与 R3 之间线缆，路由器 R1 将重新选举为主用设备 Master，R2 将选举为备用设备 Backup。

② 重新选举会导致短暂超时，假如重新连接 R1 与 R3 之间线缆，如不想要 R1 重新抢占为主用设备 Master，可关闭 R1 自动抢占功能，以下为脚本。

```
[R1-GigabitEthernet0/0/0]vrrp vrid 1 preempt-mode disable   //preempt-mode:抢占模式
```

【任务总结】

(1) VRRP 协议优先级范围＜0～255＞，但可以手动配置的范围是＜1～254＞。其中 0 被系统保留，255 保留给 IP 地址拥有者使用，无法强行指定；如果一个 VRRP 路由器将虚拟路由器的 IP 地址作为真实的接口地址，则该设备是 IP 地址拥有者，其优先级自动设为 255，无法手动修改。

(2) VRRP 自动抢占功能默认为开启状态，故障恢复后为避免业务短暂中断，可以手动关闭。在某些情况下，如主用设备间歇性宕机重启，建议关闭其 VRRP 自动抢占功能，否则周期性抢占会影响网络连通性。

(3) 当有多个路由器组建 VRRP 时，为避免未授权路由器接入并选举为主用设备，从而诱导流量并产生安全问题，管理员可对 VRRP 组成员建立认证机制，以下为脚本。

```
[R1]interface GigabitEthernet 0/0/0
[R1-GigabitEthernet0/0/0]vrrp vrid 1 authentication-mode md5 gdcp  //gdcp 为密钥
-----------------------------------------------------------------------------------
[R2]interface GigabitEthernet 0/0/0
[R2-GigabitEthernet0/0/0]vrrp vrid 1 authentication-mode md5 gdcp  //与 R1 密钥必须
                                                                     相同，否则无法
                                                                     通过对方认证
```

(4) 由于 IP 地址拥有者设备优先级无法人为修改，因此不能执行 track 命令减少其优先值。假如 R1 是 IP 地址拥有者(接口 IP 与 VRRP 组 IP 相同)，如输入以下脚本则系统会提示错误。

```
[R1]interface GigabitEthernet 0/0/0
[R1-GigabitEthernet0/0/0] vrrp  vrid  1  track  interface  GigabitEthernet  0/0/1
reduced 30
          Error: The IP address owner cannot perform the track function.
```

工作任务三

负载均衡型 VRRP

【工作目的】

掌握 VRRP 工作原理和主备备份型 VRRP 配置过程。

【工作任务】

公司购买了两个公网 IP,采用双链路接入 Internet,组建主备备份型 VRRP 保证公网接入的可靠性。随着业务发展,公司划分为技术部和工程部,单链路主备模式无法满足出口带宽需求。公司考虑切换成负载均衡型 VRRP,对于不同的 VRRP 分组,R1 与 R2 都作为主用设备,又互为备份,以充分利用现有双链路带宽资源,具体任务如下。

(1) 建立两个 VRRP 分组。在 VRRP 组 1 中,R1 作为主用设备,R2 作为备用设备;在 VRRP 组 2 中,R2 作为主用设备,R1 作为备用设备。

(2) 在 VRRP 负载均衡模式下,工程部主机 1 经 ISP-A 通往公网服务器,技术部主机 2 经 ISP-B 通往公网服务器。

(3) 验证 VRRP 协议可靠性和主备切换过程,当 ISP-A 链路故障时,主机 1 和主机 2 都经 ISP-B 通往公网服务器。

【工作背景】

在主备备份型 VRRP 中,主用设备承担所有网络流量转发业务,备用设备处于空闲状态,导致备用链路带宽资源浪费。为解决这一问题,可将 VRRP 配置为负载均衡(负载分担)模式。在该模式下,两台路由器均为主用设备,共同处理业务流量,同时又作为另一台设备的备用设备。当其中一台路由器故障后,另一台设备负责处理全部业务流量,从而保证新发起的会话能正常建立。

【任务分析】

在主备模式下,VRRP 建立单备份组,只能有一个主用设备,多个备用设备,备用设备链路处于空闲状态。

在负载均衡模式下,VRRP 建立多备份组,单个路由器同时加入不同的分组,并拥有不同的优先级。例如,在 VRRP 组 1 中,R1 作为主用设备,R2 作为备用设备;在 VRRP 组 2 中,R2 作为主用设备,R1 作为备用设备,从而两台路由器均为主用设备,同时又作为另一台设备的备用设备。内网主机可以使用不同 VRRP 分组 IP 作为默认网关,达到流量均衡的目的。

VRRP 通告报文使用 IP 多播数据包进行封装,组地址为 224.0.0.18。VRRP 协议优先级范围＜0～255＞(其中 0 被系统保留,255 保留给 IP 地址拥有者使用,无法强行指定),默认值为 100,数值越大优先级越高。

【设备器材】

汇聚层交换机(S5700)1台,路由器(AR1220)3台,需添加1GEC接口。

主机3台,承担角色见表3-1。

表3-1 主机配置表

角 色	接入方式	IP地址	网 关
主机1	eNSP PC接入	192.168.1.100/24	192.168.1.253
主机2	eNSP PC接入	192.168.1.200/24	192.168.1.254
公网服务器	eNSP Server接入	112.32.32.200/24	112.32.32.1

【环境拓扑】

环境拓扑如图3-1所示。

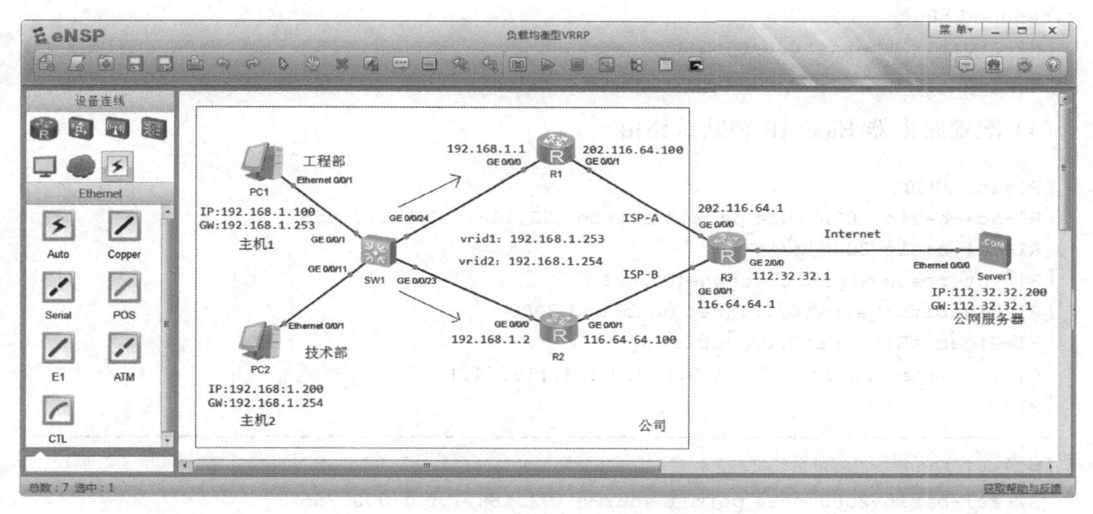

图3-1 环境拓扑

【工作过程】

1. 基本配置

(1) 路由器接口IP配置。

请读者自行根据工作拓扑图配置路由器接口IP,注意所有接口IP子网掩码均为24位。

(2) 配置路由器VRRP协议。

```
[R1]interface GigabitEthernet 0/0/0
[R1-GigabitEthernet0/0/0]vrrp vrid 1 virtual-ip 192.168.1.253
[R1-GigabitEthernet0/0/0]vrrp vrid 2 virtual-ip 192.168.1.254
[R1-GigabitEthernet0/0/0]vrrp vrid 1 priority 120     //在VRRP组1中,R1为主用设备
[R1-GigabitEthernet0/0/0]quit
[R1]
```

```
[R2]interface GigabitEthernet 0/0/0
[R2-GigabitEthernet0/0/0]vrrp vrid 1 virtual-ip 192.168.1.253
[R2-GigabitEthernet0/0/0]vrrp vrid 2 virtual-ip 192.168.1.254
[R2-GigabitEthernet0/0/0]vrrp vrid 2 priority 120     //在 VRRP 组 2 中,R2 为主用设备
[R2-GigabitEthernet0/0/0]quit
[R2]
```

(3) 在 R1 和 R2 配置上行接口监视。

```
[R1]interface GigabitEthernet 0/0/0
[R1-GigabitEthernet0/0/0]vrrp vrid 1 track interface GigabitEthernet 0/0/1 reduced 50
[R1-GigabitEthernet0/0/0]quit
[R1]
------------------------------------------------------------------------------------
[R2]interface GigabitEthernet 0/0/0
[R2-GigabitEthernet0/0/0]vrrp vrid 2 track interface GigabitEthernet 0/0/1 reduced 50
[R2-GigabitEthernet0/0/0]quit
[R2]
```

(4) 配置路由器 Easy-IP 和默认路由。

```
[R1]acl 2000
[R1-acl-basic-2000]rule permit source 192.168.1.0 0.0.0.255
[R1-acl-basic-2000]quit
[R1]interface GigabitEthernet 0/0/1
[R1-GigabitEthernet0/0/1]nat outbound 2000
[R1-GigabitEthernet0/0/1]quit
[R1]ip route-static 0.0.0.0 0.0.0.0 202.116.64.1
[R1]
------------------------------------------------------------------------------------
[R2]acl 2000
[R2-acl-basic-2000]rule permit source 192.168.1.0 0.0.0.255
[R2-acl-basic-2000]quit
[R2]interface GigabitEthernet 0/0/1
[R2-GigabitEthernet0/0/1]nat outbound 2000
[R2-GigabitEthernet0/0/1]quit
[R2]ip route-static 0.0.0.0 0.0.0.0 116.64.64.1
[R2]
```

2. 任务验证

(1) VRRP 负载均衡模式验证。

在路由器查看 VRRP 组简要信息。

```
[R1]display vrrp brief
Total:2      Master:1      Backup:1      Non-active:0
VRID  State         Interface                Type     Virtual IP
------------------------------------------------------------------------------------
1     Master        GE0/0/0                  Normal   192.168.1.253
2     Backup        GE0/0/0                  Normal   192.168.1.254
```

可以看到，在 VRRP 组 1 中，R1 是主用设备；在 VRRP 组 2 中，R1 是备用设备。

```
[R2]display vrrp brief
Total:2       Master:1       Backup:1     Non-active:0
VRID  State          Interface                  Type      Virtual IP
---------------------------------------------------------------------------
1     Backup         GE0/0/0                    Normal    192.168.1.253
2     Master         GE0/0/0                    Normal    192.168.1.254
```

可以看到，在 VRRP 组 1 中，R2 是备用设备；在 VRRP 组 2 中，R2 是主用设备。

(2) 与公网服务器的连通性验证。

在主机 1 或主机 2 的命令行窗口运行 ping 命令，TTL 值为 253，途经 2 个路由器转发，连通性验证如图 3-2 所示。

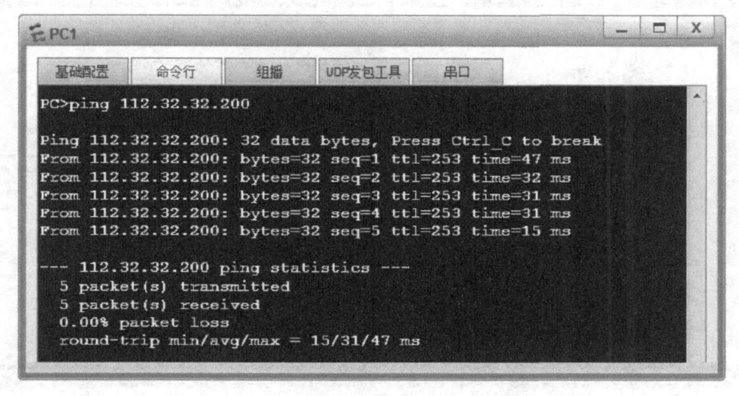

图 3-2 主机 1 可以连通公网服务器

(3) 访问公网服务器时，数据包转发路径测试。

在工程部主机 1 中的命令行窗口运行 tracert 命令，可以验证与外网通信途经 R1 转发(走 ISP-A)，如图 3-3 所示。

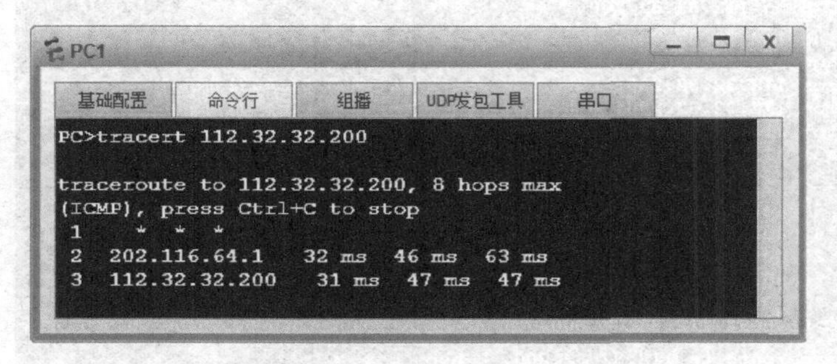

图 3-3 主机 1 数据包经 R1 转发

在技术部主机 2 的命令行窗口运行 tracert 命令，可以验证与外网通信途经 R2 转发(走 ISP-B)，如图 3-4 所示。

(4) 验证 VRRP 主备切换。

拔除 R1 与 R3 之间线缆，拓扑图如图 3-5 所示。主机 1 或主机 2 仍可以连通公网服务器，运行 tracert 命令结果如图 3-6 所示，可以验证所有主机与外网通信途经 R2 转发(走 ISP-B)。

(5) 主备切换后验证路由器 VRRP 组详细状态。

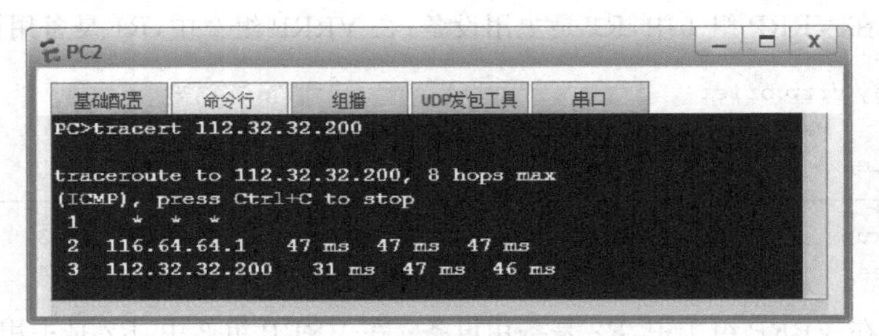

图 3-4 主机 2 数据包经 R2 转发

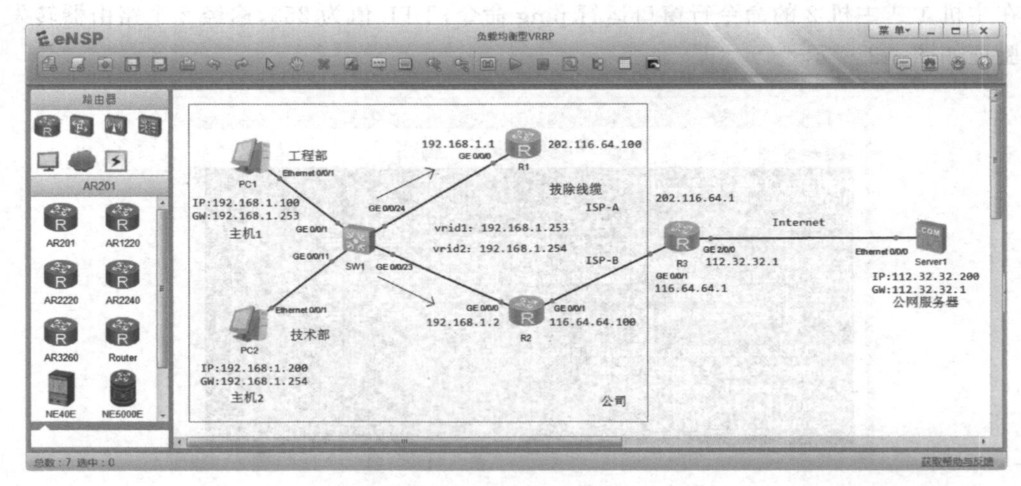

图 3-5 拔除 R1 与 R3 之间线缆

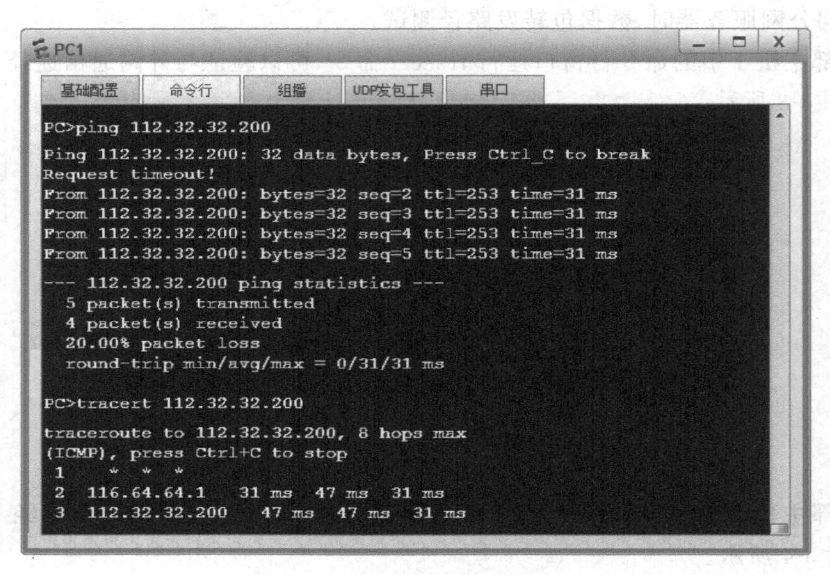

图 3-6 主机 1 数据包经 R2 转发

```
[R1]display vrrp
  GigabitEthernet0/0/0 | Virtual Router 1
    State : Backup
```

```
    Virtual IP : 192.168.1.253
    Master IP : 192.168.1.2
    PriorityRun : 70                    //优先级为 120-50=70
    PriorityConfig : 120
    MasterPriority : 100
    Preempt : YES   Delay Time : 0 s
    TimerRun : 1 s
    TimerConfig : 1 s
    Auth type : NONE
    Virtual MAC : 0000-5e00-0101
    Check TTL : YES
    Config type : normal-vrrp
    Backup-forward : disabled
    Track IF : GigabitEthernet0/0/1   Priority reduced : 50
    IF state : DOWN
    Create time : 2021-07-20 16:39:55 UTC-08:00
    Last change time : 2021-07-20 17:12:03 UTC-08:00

  GigabitEthernet0/0/0 | Virtual Router 2
    State : Backup
    Virtual IP : 192.168.1.254
    Master IP : 192.168.1.2
    PriorityRun : 100
    PriorityConfig : 100
    MasterPriority : 120
    Preempt : YES   Delay Time : 0 s
    TimerRun : 1 s
    TimerConfig : 1 s
    Auth type : NONE
    Virtual MAC : 0000-5e00-0102
    Check TTL : YES
    Config type : normal-vrrp
    Backup-forward : disabled
    Create time : 2021-07-20 16:40:00 UTC-08:00
    Last change time : 2021-07-20 16:41:23 UTC-08:00

[R2]display vrrp
  GigabitEthernet0/0/0 | Virtual Router 1
    State : Master
    Virtual IP : 192.168.1.253
    Master IP : 192.168.1.2
    PriorityRun : 100                   //100>70,R2 选举为主用设备
    PriorityConfig : 100
    MasterPriority : 100
    Preempt : YES   Delay Time : 0 s
    TimerRun : 1 s
    TimerConfig : 1 s
    Auth type : NONE
    Virtual MAC : 0000-5e00-0101
    Check TTL : YES
    Config type : normal-vrrp
    Backup-forward : disabled
```

```
    Create time : 2021-07-20 16:40:32 UTC-08:00
    Last change time : 2021-07-20 17:12:03 UTC-08:00

  GigabitEthernet0/0/0 | Virtual Router 2
    State : Master
    Virtual IP : 192.168.1.254
    Master IP : 192.168.1.2
    PriorityRun : 120                    //120>100,R2 选举为主用设备
    PriorityConfig : 120
    MasterPriority : 120
    Preempt : YES   Delay Time : 0 s
    TimerRun : 1 s
    TimerConfig : 1 s
    Auth type : NONE
    Virtual MAC : 0000-5e00-0102
    Check TTL : YES
    Config type : normal-vrrp
    Backup-forward : disabled
    Track IF : GigabitEthernet0/0/1  Priority reduced : 50
    IF state : UP
    Create time : 2021-07-20 16:40:37 UTC-08:00
    Last change time : 2021-07-20 16:41:23 UTC-08:00
```

【任务总结】

(1) 将 VRRP 主备模式切换为负载均衡模式时，需要对 IP 重新规划，注意 VRRP 所有组 IP 要与接口(该接口已加入 VRRP 组)IP 处于同一网段，否则系统会提示出错；

(2) 在规划 IP 时，由于 IP 地址拥有者设备优先级无法人为修改，也不能执行 track 命令减少其优先值，因此不建议将路由器接口 IP 与 VRRP 组 IP 设为相同，否则如拔除 R1 与 R3 之间线缆，由于在 VRRP 组 1 中 R1 仍为主用设备，故会导致主机 1 无法连通公网。

工作任务四

MSTP+VRRP 实现网关冗余和负载均衡

【工作目的】

理解 MSTP 的工作原理,掌握配置 MSTP 实现负载均衡过程。

【工作任务】

A 公司(包含技术部和工程部)收购 B 公司(包含销售部和财务部)。为提高公司合并后内网连通的高带宽和可靠性,公司决定采购两台三层交换机 SW3 和 SW4 以提供链路冗余和负载均衡,需要管理员同时配置 MSTP、链路聚合与 VRRP 三种协议,具体任务如下。

(1) 配置 MSTP 实现负载均衡。在 SW1,Vlan10 流量走 SW1-SW3 之间链路,Vlan20 走 SW1-SW4 之间链路;在 SW2,Vlan30 流量走 SW2-SW3 之间链路,Vlan40 走 SW2-SW4 之间链路。

(2) 在三层交换机配置链路聚合,通过负载分担方式提高三层交换机之间链路带宽。

(3) 在三层交换机配置 VRRP 实现负载均衡。对于 Vlan10 和 Vlan30,SW3 是主用设备;对于 Vlan20 和 Vlan40,SW4 是主用设备。

(4) 验证 MSTP、链路聚合与 VRRP 三种协议主备切换过程,当备份链路故障时,自动切换到主链路。

【工作背景】

利用 MSTP、链路聚合与 VRRP 都能达到冗余备份和负载均衡的目的。如同时使用三种协议组建局域网,则可以有效提高内网互联的可靠性。

【设备器材】

接入层交换机(S3700)2 台,汇聚层交换机(S5700)2 台,路由器(AR1220)2 台,需要添加 2SA 两个串口模块。

主机 4 台,承担角色见表 4-1。

表 4-1 主机配置表

角　色	接入方式	IP 地址	Vlan ID
主机 1	eNSP PC 接入	192.168.10.10/24	Vlan10(技术部)
主机 2	eNSP PC 接入	192.168.20.10/24	Vlan20(工程部)
主机 3	eNSP PC 接入	192.168.30.20/24	Vlan30(销售部)
主机 4	eNSP PC 接入	192.168.40.20/24	Vlan40(财务部)

【环境拓扑】

环境拓扑如图 4-1 所示。

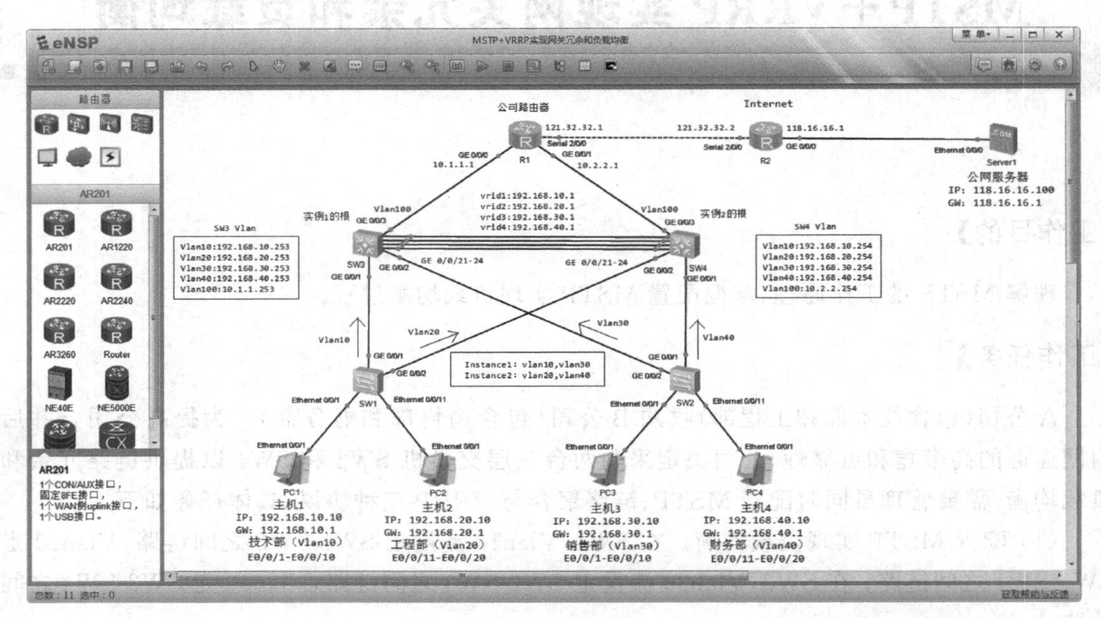

图 4-1　环境拓扑

【工作过程】

1. 基本配置

(1) 交换机 Vlan 和端口配置。

```
<Huawei>system-view
[Huawei]sysname SW1
[SW1]vlan batch 10 20
[SW1]port-group 1                                    //技术部组
[SW1-port-group-1]group-member Ethernet 0/0/1 to Ethernet 0/0/10
[SW1-port-group-1]port link-type access
[SW1-port-group-1]port default vlan 10
[SW1-port-group-1]quit
[SW1]port-group 2                                    //工程部组
[SW1-port-group-2]group-member Ethernet 0/0/11 to Ethernet 0/0/20
[SW1-port-group-2]port link-type access
[SW1-port-group-2]port default vlan 20
[SW1-port-group-2]quit
[SW1]port-group 3                                    //Trunk 组
[SW1-port-group-3]group-member GigabitEthernet 0/0/1 GigabitEthernet 0/0/2
[SW1-port-group-3]port link-type trunk
[SW1-port-group-3]port trunk allow-pass vlan all
[SW1-port-group-3]quit
[SW1]
---------------------------------------------------------------------------
```

```
<Huawei>system-view
[Huawei]sysname SW2
[SW2]vlan batch 30 40
[SW2]port-group 1
[SW2-port-group-1]group-member Ethernet 0/0/1 to Ethernet 0/0/10      //销售部组
[SW2-port-group-1]port link-type access
[SW2-port-group-1]port default vlan 30
[SW2-port-group-1]quit
[SW2]port-group 2
[SW2-port-group-2]group-member Ethernet 0/0/11 to Ethernet 0/0/20     //财务部组
[SW2-port-group-2]port link-type access
[SW2-port-group-2]port default vlan 40
[SW2-port-group-2]quit
[SW2]port-group 3                                                      //Trunk 组
[SW2-port-group-3]group-member GigabitEthernet 0/0/1 GigabitEthernet 0/0/2
[SW2-port-group-3]port link-type trunk
[SW2-port-group-3]port trunk allow-pass vlan all
[SW2-port-group-3]quit
[SW2]
-----------------------------------------------------------------------------------
<Huawei>system-view
[Huawei]sysname SW3
[SW3]vlan batch 10 20 30 40 100
[SW3]port-group 1                                                      //对内 Trunk 组
[SW3-port-group-1]group-member GigabitEthernet 0/0/1 to GigabitEthernet 0/0/2
[SW3-port-group-1]port link-type trunk
[SW3-port-group-1]port trunk allow-pass vlan all
[SW3-port-group-1]quit
[SW3]interface GigabitEthernet 0/0/3
[SW3-GigabitEthernet0/0/3]port link-type trunk
[SW3-GigabitEthernet0/0/3]port trunk allow-pass vlan all   //如果没有本行命令,SW3 的
                                                             Vlan100 接口无法 ping 通
                                                             R1 的 G0/0/0 接口
[SW3-GigabitEthernet0/0/3]port trunk pvid vlan 100   //将端口更改为默认 Vlan,Access 模式
                                                       为"port default vlan100",Trunk
                                                       模式为"port trunk pvid vlan 40"。
                                                       Vlan100 是虚拟接口,UP 的前提条件
                                                       首先是要有 IP 地址,其次是 Vlan100
                                                       里面有物理接口,否则为 Down
[SW3-GigabitEthernet0/0/3]quit
[SW3]interface Vlanif 10
[SW3-Vlanif10]ip address 192.168.10.253 24
[SW3-Vlanif10]interface Vlanif 20
[SW3-Vlanif20]ip address 192.168.20.253 24
[SW3-Vlanif20]interface Vlanif 30
[SW3-Vlanif30]ip address 192.168.30.253 24
[SW3-Vlanif30]interface Vlanif 40
[SW3-Vlanif40]ip address 192.168.40.253 24
[SW3-Vlanif40]interface Vlanif 100
[SW3-Vlanif100]ip address 10.1.1.253 24
[SW3-Vlanif100]quit
[SW3]
-----------------------------------------------------------------------------------
<Huawei>system-view
[Huawei]sysname SW3
```

```
[SW4]vlan batch 10 20 30 40 100
[SW4]port-group 1                                          //对内 Trunk 组
[SW4-port-group-1]group-member GigabitEthernet 0/0/1 to GigabitEthernet 0/0/2
[SW4-port-group-1]port link-type trunk
[SW4-port-group-1]port trunk allow-pass vlan all
[SW4-port-group-1]quit
[SW4]interface GigabitEthernet 0/0/3
[SW4-GigabitEthernet0/0/3]port link-type trunk
[SW4-GigabitEthernet0/0/3]port trunk allow-pass vlan all
[SW4-GigabitEthernet0/0/3]port trunk pvid vlan 100
[SW4-GigabitEthernet0/0/3]quit
[SW4]interface Vlanif 10
[SW4-Vlanif10]ip address 192.168.10.254 24
[SW4-Vlanif10]interface Vlanif 20
[SW4-Vlanif20]ip address 192.168.20.254 24
[SW4-Vlanif20]interface Vlanif 30
[SW4-Vlanif30]ip address 192.168.30.254 24
[SW4-Vlanif30]interface Vlanif 40
[SW4-Vlanif40]ip address 192.168.40.254 24
[SW4-Vlanif40]interface Vlanif 100
[SW4-Vlanif100]ip address 10.2.2.254 24
[SW4-Vlanif100]quit
[SW4]
```

(2) 配置链路聚合。

```
[SW3]interface Eth-Trunk 1                        //创建以太网 Trunk 链路(链路聚合组)1
[SW3-Eth-Trunk1]mode manual load-balance          //manual load-balance:手动负载分担方式;
                                                    lacp-static:静态 LACP 模式,通过 LACP
                                                    模式协议协商参数自动选择接口运行模式
```

注意:

① 在 manual load-balance 模式中,将成员接口加入链路聚合组由手动配置。该模式下所有活动链路都参与数据的转发,平均分担流量。

② 在 lacp-static 模式中,可以实现链路负载分担和链路冗余备份双重功能,也称为 M : N 模式。在链路聚合组中 M 条(可人为指定)链路处于活动状态,负责转发数据并进行负载分担,剩下 N 条链路处于非活动状态作为备份链路,不转发数据。当 M 条链路中有链路出现故障时,系统会从 N 条备份链路中选择优先级最高的接替出现故障的链路,并开始转发数据。

③ 如选择 lacp-static 模式,需另外执行[SW3-Eth-Trunk1]bpdu enable 命令,交换机才能接收并处理 LACP 协议报文,否则直接丢弃,Eth-Trunk 链路状态为 Down。

```
[SW3-Eth-Trunk1]trunkport GigabitEthernet 0/0/21 to 0/0/24  //将 G0/0/21- G0/0/24 接
                                                              口加入 Eth-Trunk 1 链
                                                              路聚合组 1
[SW3-Eth-Trunk1]port link-type trunk              //注意 trunk 属性需在链路聚合组中配置,不能
                                                    在原物理接口(G0/0/21- G0/0/24)中配置
[SW3-Eth-Trunk1]port trunk allow-pass vlan all
[SW3-Eth-Trunk1]quit
[SW3]
------------------------------------------------------------------------------------------
[SW4]int Eth-Trunk 1
[SW4-Eth-Trunk1]mode manual load-balance
```

```
[SW4-Eth-Trunk1]trunkport GigabitEthernet 0/0/21 to 0/0/24
[SW4-Eth-Trunk1]port link-type trunk
[SW4-Eth-Trunk1]port trunk allow-pass vlan all
[SW4-Eth-Trunk1]quit
[SW4]
```

注意：思科配置端口聚合脚本如下(以下不需要配置)。

```
Switch(config)#interface range fastEthernet 0/21-24
Switch(config-if-range)#channel-group 1 mode active
Switch(config-if-range)#exit
Switch(config)#interface port-channel 1
Switch(config-if)#switchport trunk encapsulation dot1q
Switch(config-if-range)#switchport mode trunk
```

(3) 创建 MSTP 实例与 Vlan 映射关系。

```
[SW1]stp mode mstp
[SW1]stp region-configuration
[SW1-mst-region]region-name gdcp_company
[SW1-mst-region]revision-level 1
[SW1-mst-region]instance 1 vlan 10 30
[SW1-mst-region]instance 2 vlan 20 40
[SW1-mst-region]active region-configuration
[SW1-mst-region]quit
[SW1]
--------------------------------------------------------------------------------
[SW2]stp mode mstp
[SW2]stp region-configuration
[SW2-mst-region]region-name gdcp_company
[SW2-mst-region]revision-level 1
[SW2-mst-region]instance 1 vlan 10 30
[SW2-mst-region]instance 2 vlan 20 40
[SW2-mst-region]active region-configuration
[SW2-mst-region]quit
[SW2]
--------------------------------------------------------------------------------
[SW3]stp mode mstp
[SW3]stp region-configuration
[SW3-mst-region]region-name gdcp_company
[SW3-mst-region]revision-level 1
[SW3-mst-region]instance 1 vlan 10 30
[SW3-mst-region]instance 2 vlan 20 40
[SW3-mst-region]active region-configuration
[SW3-mst-region]quit
[SW3]
--------------------------------------------------------------------------------
[SW4]stp mode mstp
[SW4]stp region-configuration
[SW4-mst-region]region-name gdcp_company
[SW4-mst-region]revision-level 1
[SW4-mst-region]instance 1 vlan 10 30
[SW4-mst-region]instance 2 vlan 20 40
[SW4-mst-region]active region-configuration
```

```
[SW4-mst-region]quit
[SW4]
```

(4) 指定不同实例 MST 根交换机。

```
[SW3]stp instance 1 root primary          //禁止实例 1 走 SW1-SW4 链路,和 SW2-SW4 链路
[SW4]stp instance 2 root primary          //禁止实例 2 走 SW1-SW3 链路,和 SW2-SW3 链路
```

(5) 在三层交换机 Vlan 中建立 VRRP 组,指定主用设备和备用设备。

```
[SW3]int vlan10
[SW3-Vlanif10]vrrp vrid 1 virtual-ip 192.168.10.1  //注意,VRRP 组 IP 在 Vlan 中配置,以
                                                     实现 VRRP 负载均衡。其中 Vlan 10
                                                     和 Vlan 30 的主用设备为 SW3,Vlan
                                                     20 和 Vlan 40 的主用设备为 SW4
[SW3-Vlanif10]vrrp vrid 1 priority 120         //对于 Vlan 10 为主用设备
[SW3-Vlanif10]vrrp vrid 1 track interface GigabitEthernet 0/0/3 reduced 30
[SW3-Vlanif10]int vlan 20
[SW3-Vlanif20]vrrp vrid 2 virtual-ip 192.168.20.1
[SW3-Vlanif20]int vlan 30
[SW3-Vlanif30]vrrp vrid 3 virtual-ip 192.168.30.1
[SW3-Vlanif30]vrrp vrid 3 priority 120          //对于 Vlan 30 为主用设备
[SW3-Vlanif30]vrrp vrid 3 track interface GigabitEthernet 0/0/3 reduced 30
[SW3-Vlanif30]int vlan 40
[SW3-Vlanif40]vrrp vrid 4 virtual-ip 192.168.40.1
[SW3-Vlanif40]quit
[SW3]
--------------------------------------------------------------------------------
[SW4]int vlan10
[SW4-Vlanif10]vrrp vrid 1 virtual-ip 192.168.10.1
[SW4-Vlanif10]int vlan 20
[SW4-Vlanif20]vrrp vrid 2 virtual-ip 192.168.20.1
[SW4-Vlanif20]vrrp vrid 2 priority 120          //对于 Vlan 20 为主用设备
[SW4-Vlanif20]vrrp vrid 2 track interface GigabitEthernet 0/0/3 reduced 30
[SW4-Vlanif20]int vlan 30
[SW4-Vlanif30]vrrp vrid 3 virtual-ip 192.168.30.1
[SW4-Vlanif30]int vlan 40
[SW4-Vlanif40]vrrp vrid 4 virtual-ip 192.168.40.1
[SW4-Vlanif40]vrrp vrid 4 priority 120          //对于 Vlan 40 为主用设备
[SW4-Vlanif40]vrrp vrid 4 track interface GigabitEthernet 0/0/3 reduced 30
[SW4-Vlanif40]quit
[SW4]
```

(6) 在三层交换机配置 OSPF 路由和默认路由。

```
[SW3]ospf 1
[SW3-ospf-1]area 0
[SW3-ospf-1-area-0.0.0.0]network 192.168.10.0 0.0.0.255
[SW3-ospf-1-area-0.0.0.0]network 192.168.20.0 0.0.0.255
[SW3-ospf-1-area-0.0.0.0]network 192.168.30.0 0.0.0.255
[SW3-ospf-1-area-0.0.0.0]network 192.168.40.0 0.0.0.255
[SW3-ospf-1-area-0.0.0.0]network 10.1.1.0 0.0.0.255     //与 R1 建立邻接关系
[SW3-ospf-1-area-0.0.0.0]quit
[SW3-ospf-1]quit
```

```
[SW3]ip route-static 0.0.0.0 0.0.0.0 10.1.1.1
[SW3]
--------------------------------------------------------------------------------
[SW4]ospf 1
[SW4-ospf-1]area 0
[SW4-ospf-1-area-0.0.0.0]network 192.168.10.0 0.0.0.255
[SW4-ospf-1-area-0.0.0.0]network 192.168.20.0 0.0.0.255
[SW4-ospf-1-area-0.0.0.0]network 192.168.30.0 0.0.0.255
[SW4-ospf-1-area-0.0.0.0]network 192.168.40.0 0.0.0.255
[SW4-ospf-1-area-0.0.0.0]network 10.2.2.0 0.0.0.255
[SW4-ospf-1-area-0.0.0.0]quit
[SW4-ospf-1]quit
[SW4]ip route-static 0.0.0.0 0.0.0.0 10.2.2.1
[SW4]
```

(7) 在路由器上配置 OSPF 路由、默认路由和 Easy-IP。

请读者自行根据工作拓扑图配置路由器 R1 和 R2 接口 IP,注意所有接口子网掩码均为 24 位。

```
[R1]ospf 1
[R1-ospf-1]area 0
[R1-ospf-1-area-0.0.0.0]network 10.1.1.0 0.0.0.255
[R1-ospf-1-area-0.0.0.0]network 10.2.2.0 0.0.0.255
[R1-ospf-1-area-0.0.0.0]quit
[R1-ospf-1]quit
[R1]ip route-static 0.0.0.0 0.0.0.0 121.32.32.2
[R1]acl 2000
[R1-acl-basic-2000]rule permit source any
[R1-acl-basic-2000]quit
[R1]interface Serial 2/0/0
[R1-Serial2/0/0]nat outbound 2000
[R1-Serial2/0/0]quit
[R1]
```

2. 任务验证

(1) VRRP 负载均衡验证。

在三层交换机上查看 VRRP 组简要信息。

```
[SW3]display vrrp brief
VRID  State       Interface          Type     Virtual IP
--------------------------------------------------------------------------------
1     Master      Vlanif10           Normal   192.168.10.1
2     Backup      Vlanif20           Normal   192.168.20.1
3     Master      Vlanif30           Normal   192.168.30.1
4     Backup      Vlanif40           Normal   192.168.40.1
--------------------------------------------------------------------------------
```

对于 Vlan10 和 Vlan30,SW3 为主用设备。

```
[SW4]display vrrp brief
VRID  State       Interface          Type     Virtual IP
--------------------------------------------------------------------------------
1     Backup      Vlanif10           Normal   192.168.10.1
```

```
2      Master        Vlanif20                Normal   192.168.20.1
3      Backup        Vlanif30                Normal   192.168.30.1
4      Master        Vlanif40                Normal   192.168.40.1
--------------------------------------------------------------------------------
Total:4      Master:2      Backup:2      Non-active:0
```

对于 Vlan20 和 Vlan40，SW4 为主用设备。

(2) 内网主机的连通性验证与链路聚合冗余验证。

在主机 1 的命令行窗口输入 ping 192.168.40.10 -t，连通后拔除 SW3 和 SW4 之间任意 3 条线缆，如图 4-2 所示。发现主机 1 丢包后继续 ping 通主机 4，如图 4-3 所示，从而验证利用链路聚合达到冗余备份的目的。

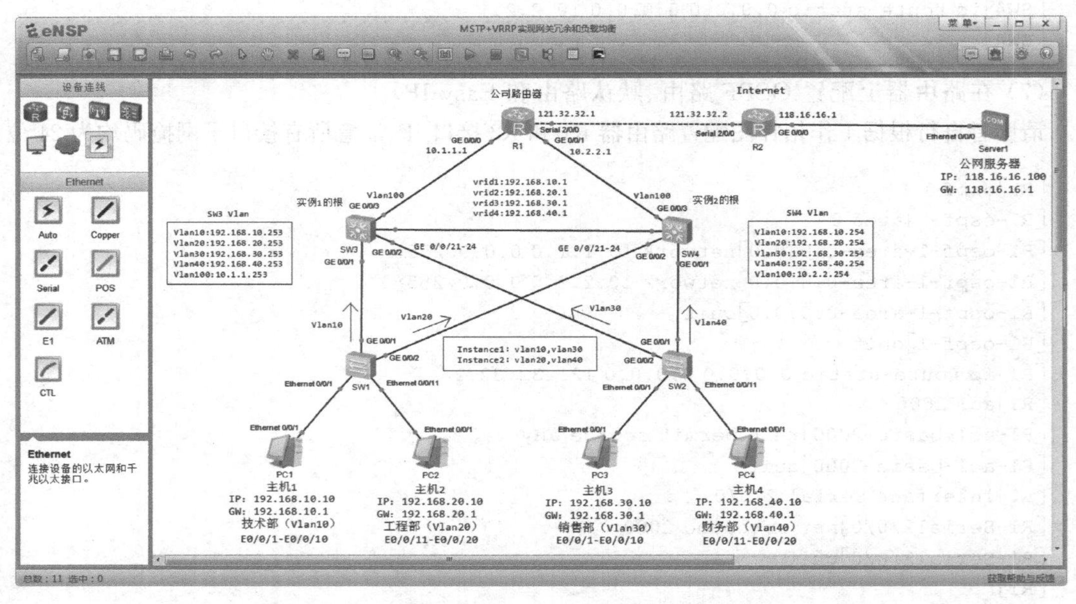

图 4-2　拔除 SW3 和 SW4 之间任意 3 条线缆

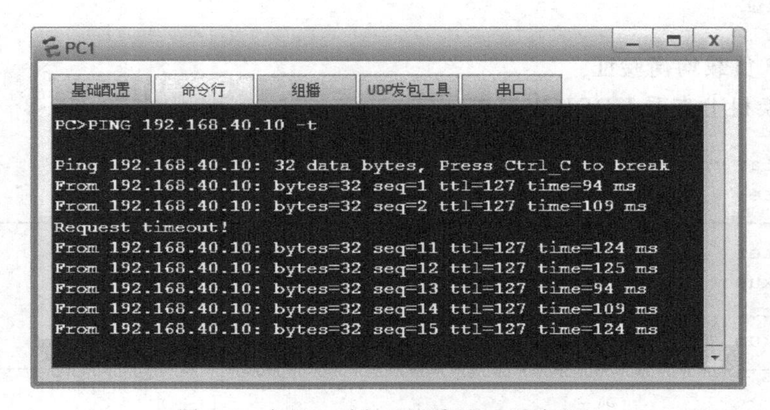

图 4-3　主机 1 丢包后继续 ping 通主机 4

(3) 公网服务器连通性验证与 MSTP 负载均衡验证。

所有内网主机可以连通公网服务器。其中主机 1 和主机 3 通过 SW3(IP 地址：192.168.10.253/192.168.30.253)连通公网服务器，如图 4-4 所示。

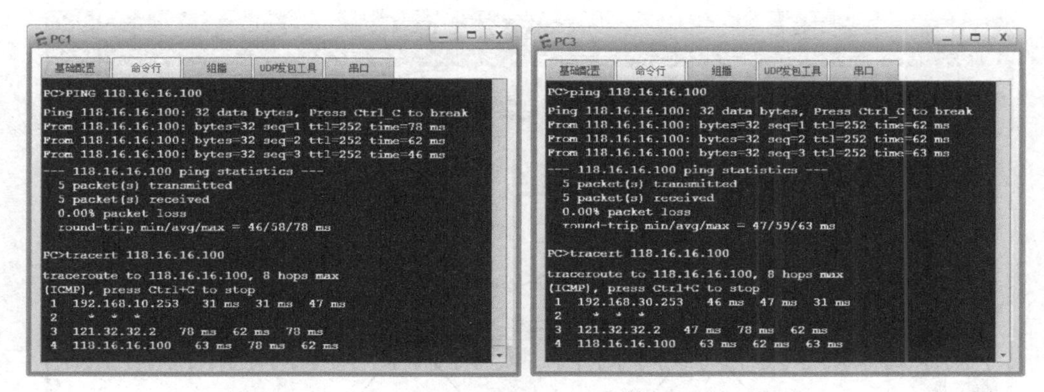

图 4-4　主机 1 和主机 3 通过 SW3 连通公网服务器

主机 2 和主机 4 通过 SW4(IP 地址：192.168.20.254/192.168.40.254)连通公网服务器,如图 4-5 所示。从而验证利用 MSTP 实现负载均衡的目的。

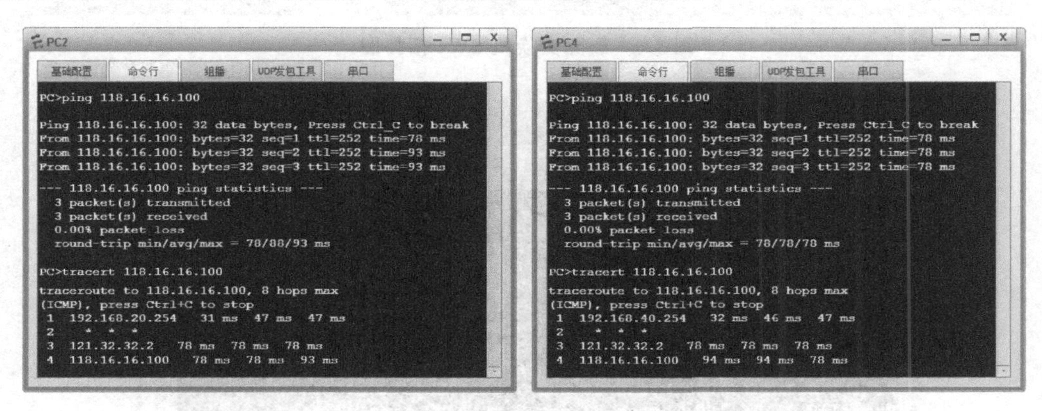

图 4-5　主机 2 和主机 4 通过 SW4 连通公网服务器

(4) 验证 VRRP 主备切换。

在主机 1 和主机 2 的命令行窗口中输入 ping 118.16.16.100 -t,连通后删除交换机 SW4,如图 4-6 所示。发现主机 1 仍可连通公网服务器,期间并未发生丢包现象,如图 4-7 所示,这是因为对于 Vlan10,SW3 是主用设备,删除交换机 SW4 不受任何影响。主机 2 丢若干个包后继续连通公网服务器,如图 4-8 所示,这是因为对于 Vlan20,SW4 是主用设备,从主用设备 SW4 切换到备用设备 SW3 需要等待选举时长,导致丢包现象。以上验证了利用 VRRP 达到冗余备份的目的。

在 SW3 查看 VRRP 组简要信息。

```
[SW3]display vrrp brief
VRID  State       Interface           Type     Virtual IP
----------------------------------------------------------------------------
1     Master      Vlanif10            Normal   192.168.10.1
2     Master      Vlanif20            Normal   192.168.20.1
3     Master      Vlanif30            Normal   192.168.30.1
4     Master      Vlanif40            Normal   192.168.40.1
----------------------------------------------------------------------------
Total:4     Master:4     Backup:0     Non-active:0
```

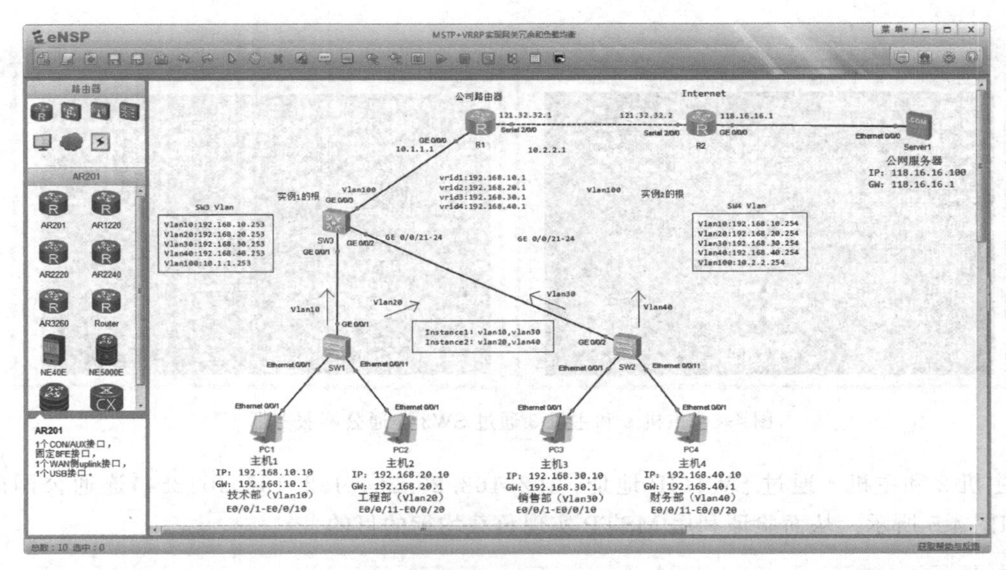

图 4-6　删除交换机 SW4 拓扑图

图 4-7　删除 SW4 后，主机 1 仍可连通公网服务器

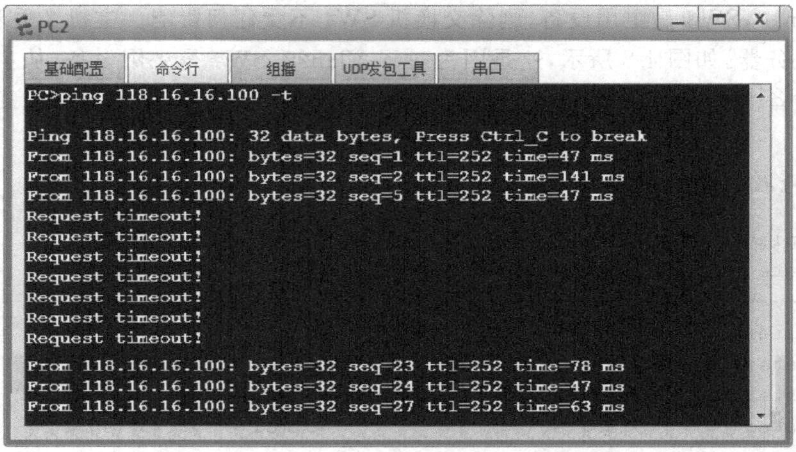

PC2

基础配置　命令行　组播　UDP发包工具　串口

```
PC>ping 118.16.16.100 -t

Ping 118.16.16.100: 32 data bytes, Press Ctrl_C to break
From 118.16.16.100: bytes=32 seq=1 ttl=252 time=47 ms
From 118.16.16.100: bytes=32 seq=2 ttl=252 time=141 ms
From 118.16.16.100: bytes=32 seq=5 ttl=252 time=47 ms
Request timeout!
Request timeout!
Request timeout!
Request timeout!
Request timeout!
Request timeout!
Request timeout!
From 118.16.16.100: bytes=32 seq=23 ttl=252 time=78 ms
From 118.16.16.100: bytes=32 seq=24 ttl=252 time=47 ms
From 118.16.16.100: bytes=32 seq=27 ttl=252 time=63 ms
```

图 4-8　删除 SW4 后，主机 2 丢包后继续连通公网服务器

可以看到，此时对于所有 Vlan，SW3 均为主用设备。

【任务总结】

（1）链路聚合如无特殊要求，建议使用 manual load-balance（手动负载分担）方式，既可增加链路带宽，又能进行链路备份。

（2）VRRP 可以在路由器上配置，也可以在三层交换机上配置。如果在路由器上配置 VRRP，目的是对出口网关进行冗余备份和负载均衡，提高公网访问的可靠性；如果在三层交换机上配置 VRRP，目的是对内网汇聚层进行冗余备份和负载均衡，提高 Vlan 之间转发的可靠性。

（3）与路由器不同，在三层交换机上配置负载均衡性 VRRP，应在 Vlan 中创建 VRRP 组。

工作任务五

多区域 OSPF 配置

【工作目的】

理解 OSPF 区域划分原理，掌握 OSPF 多区域配置和发布默认路由。

【工作任务】

A 公司收购 B 公司和 C 公司并确定上市。为标识公司区域结构，减少网络组播流量，提高合并后整体内网连通的可靠性，公司决定内部划出 3 个不同区域(公司 A 为 Area0，公司 B 为 Area10，公司 C 为 Area20)，通过 OSPF 路由协议互联，外部通过公网 IP(202.116.64.1)连接至 Internet，具体任务如下。

(1) 在 R1、R2、R3 和 R4 上配置 OSPF 协议，实现主干区域 Area0 互通。

(2) 在 R5、R6 和 R7 上配置 OSPF 协议，实现常规区域 Area10 互通。

(3) 在 R8、R9 和 R10 上配置 OSPF 协议，实现常规区域 Area20 互通。

(4) 内网主机之间可以相互连通，并能通过公网 IP(202.116.64.1)访问公网服务器。

【工作背景】

在 OSPF 单区域中，每台路由器都收集其他路由器发来的链路状态信息。如果网络规模不断扩大，链路状态信息也随之增多，使得单台路由器上链路状态数据库非常庞大，如路由器在庞大的路由表中逐一匹配路由条目会极大影响转发效率。为解决这一问题，OSPF 将整个区域划分成多个自治系统，每个自治系统成为一个区域，其中 Area0 称为骨干区域，其他区域称为非骨干区域(其他区域必须与骨干区域 Area0 相连)。划分为不同区域的目的是更好地管理，就像一个国家一样，国土面积很大，难以统一管理，则将整个国家划分成不同省份，由首都(Area0)分而治之。OSPF 多区域划分可以增强网络扩展性，降低路由器负载，实现快速收敛，适合组建大规模内网拓扑，如企业复杂内网、电信连通内网、国家电网等。

【任务分析】

1. 单区域内路由器角色和选举过程

在一个自治区域内，为减少 OSPF 流量和 CPU 消耗，将 OSPF 路由器分为 DR 指定路由器，BDR 备份指定路由器和 Dother 其他路由器。

(1) DR。监听网络中所有 LSA 状态信息，并向 BDR 和 Dother 通告所有 OSPF 网段汇总信息，可以认为是班长，收集同学意见，汇总后交给同学。

(2) BDR。作为 DR 的备份，收到 LSA 并监听 DR 状态，不允许抢占，DR 没了，DBDR 才能上，可以认为是副班长。

(3) Dother。只向 DR/BDR 发送 LSA,并接收 DR 发送的 OSPF 网段汇总信息,可以认为是普通同学。

2. 多区域间 OSPF 路由器类型

为减少路由组播和 CPU 负担,链路状态信息 LSA 只在一个区域内部泛洪,区域间仅通告汇总后的路由条目。

(1) 区域内部路由器。保存区域内部的链路状态信息。

(2) 区域边界路由器(ABR)。连接 Area0 骨干区域和其他非骨干区域的路由器,负责通告区域间 OSPF 路由;

(3) 自治边界路由器(ASBR)。连接 OSPF 区域与外部(其他非 OSPF 区域)的路由器。

3. OSPF 区域类型

(1) 骨干区域。Area0,也称为中转区域。

(2) 非骨干区域。除 Area0 之外的其他区域,也称为常规区域。为防止区域间产生环路,导致区域间路由循环通告,所有非骨干区域之间路由信息必须通过骨干区域通告,即所有非骨干区域必须和骨干区域连接。其中非骨干区域可划分为标准区域和末梢区域。

① 标准区域:禁用外部 AS 路由信息进入。

② stub 区域:末梢区域,只接收区域间路由,禁止外部 AS(其他非 OSPF 区域)进入。其中末梢区域又可分为完全末梢区域和非纯末梢区域。完全末梢区域(totally stub),禁止所有外部 AS 和区域间路由信息;非纯末梢区域(NSSA 区域),禁用非直连的外部 AS 信息进入。

4. Router ID 作业及选举规则

每台路由器必须采用唯一 id 标识自己,长度 32 位。在一个 OSPF 域内,路由器需要知道其他成员组身份,如邻居路由器、指定路由器(DR)、备份指定路由器 BDR 等,路由器身份采用 Router ID 标识。如路由器名称 R1,人能理解,但别的路由器无法理解,此时可对 R1 配置 Router ID,脚本为 ospf router-id 1.1.1.1(id 值可以自定义,一般选用 IP 地址作为其值)。如不指定 Router ID,Router ID 也可以通过选举产生,过程如下。

(1) 选取最大 Loopback 接口 IP 地址作为 Router ID。

(2) 如果没有配置 Loopback 接口 IP 地址,则选择物理接口最大 IP 地址作为 Router ID。

5. ospf 两种组播地址类型

(1) 224.0.0.5。负责通告,即发给 DR/other 的报文组播 IP 都是 224.0.0.5。

(2) 224.0.0.6。负责监听,即其他路由器发送给 DR/BDR 的组播 IP 都是 224.0.0.6。

6. ospf 建立邻接关系七种状态

(1) Down。没有启用 OSPF 进程的状态,或者邻居失效后的状态。

(2) Init。初始化状态,OSPF 路由器以固定时隙(默认 10s)发送 Hello 分组,以建立邻居(邻接)关系。路由器第一次收到对端发来的 HELLO 包时,将对端的状态设置为 Init。

(3) Two-Way。邻居状态,每台 OSPF 路由器与网络中其他路由建立双向会话以建立邻居。当路由器看到自身出现在邻居路由器发送的 HELLO 分组中时,则进入邻居双向状态。

(4) ExStart。准启动状态,交换信息的初始化状态,用于选举主从路由器,利用 HELLO 报文中的 Router ID 值选举 DR(指定路由器)和 BDR(备份指定路由器)。

(5) Exchange。交换信息状态,通告自身网段信息。

(6) Loading。加载状态,没有学习完的状态,用于向邻居学习对端网段信息。

(7) Full Adjacency。邻接状态,网段信息学习完成状态。加载状态结束后,路由器进入邻接状态,完成 LSDB 同步。

【设备器材】

路由器(AR1220)11 台,需要添加 1GEC 千兆接口模块或 2SA 串口模块。

主机 4 台,承担角色见表 5-1。

表 5-1 主机配置表

角 色	接入方式	IP 地址	所属公司
主机 1	eNSP PC 接入	192.168.5.10/24	公司 A
主机 2	eNSP PC 接入	10.0.6.10/24	公司 B
主机 3	eNSP PC 接入	172.16.6.10/24	公司 C
公网服务器	eNSP Server 接入	116.64.64.100	

【环境拓扑】

环境拓扑如图 5-1 所示。

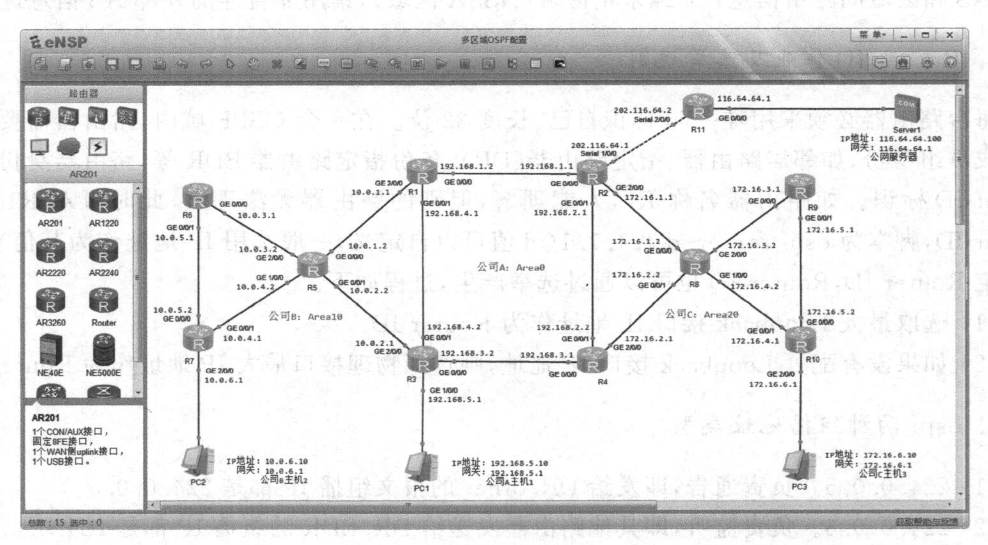

图 5-1 环境拓扑

【工作过程】

1. 基本配置

(1) 路由器接口 IP 和系统名称配置。

请读者根据网络拓扑自行配置路由器接口 IP 和系统名称,注意所有接口 IP 地址子网掩码长度均为24。

(2) 路由器 OSPF 协议配置。

```
[R1]ospf 1
[R1-ospf-1]area 0
[R1-ospf-1-area-0.0.0.0]network 192.168.1.0 0.0.0.255
[R1-ospf-1-area-0.0.0.0]network 192.168.4.0 0.0.0.255
[R1-ospf-1-area-0.0.0.0]quit
[R1-ospf-1]area 10
[R1-ospf-1-area-0.0.0.10]network 10.0.1.0 0.0.0.255
[R1-ospf-1-area-0.0.0.10]quit
[R1-ospf-1]quit
[R1]
------------------------------------------------------------------------
[R5]ospf 1
[R5-ospf-1]area 10
[R5-ospf-1-area-0.0.0.10]network 10.0.1.0 0.0.0.255
[R5-ospf-1-area-0.0.0.10]network 10.0.2.0 0.0.0.255
[R5-ospf-1-area-0.0.0.10]network 10.0.3.0 0.0.0.255
[R5-ospf-1-area-0.0.0.10]network 10.0.4.0 0.0.0.255
[R5-ospf-1-area-0.0.0.10]quit
[R5-ospf-1]quit
[R5]
------------------------------------------------------------------------
[R8]ospf 1
[R8-ospf-1]area 20
[R8-ospf-1-area-0.0.0.20]network 172.16.1.0 0.0.0.255
[R8-ospf-1-area-0.0.0.20]network 172.16.2.0 0.0.0.255
[R8-ospf-1-area-0.0.0.20]network 172.16.3.0 0.0.0.255
[R8-ospf-1-area-0.0.0.20]network 172.16.4.0 0.0.0.255
[R8-ospf-1-area-0.0.0.20]quit
[R8-ospf-1]quit
[R8]
```

请读者根据网络拓扑继续配置其他路由器 OSPF 路由。由于 OSPF 协议属于 IGP 内部网关协议，请勿宣告公网网段。如 R2 不需宣告＜202.116.64.0＞网段，R11 也不需开启 OSPF 进程，避免公网路由器发现内网路由条目。

(3) 在路由器 R2 上配置默认路由，并通过 OSPF 协议下发默认路由。

```
[R2]ip route-static 0.0.0.0 0.0.0.0 202.116.64.2
[R2]ospf 1
[R2-ospf-1]default-route-advertise  //向其他路由器通过 OSPF 协议通告一条 0.0.0.0 的默认
                                      路由，指向本地路由器 R2，使其他路由器能够连接公网
[R2-ospf-1]quit
[R2]
```

注意：

① [R2-ospf-1]default-route-advertise：向其他路由器通告本地路由器 R2 是他们的默认路由，前提是 R2 必须存在默认路由。假如 R2 没有配置默认路由，则不向其他路由器通告下发默认路由。

② [R2-ospf-1]default-route-advertise always：无论本地路由器R2是否配置默认路由，都向其他路由器通告下发默认路由。

(4) 在路由器R2上配置Easy-IP。

```
[R2]acl 2000
[R2-acl-basic-2000]rule permit source any
[R2-acl-basic-2000]quit
[R2]interface Serial 1/0/0
[R2-Serial1/0/0]nat outbound 2000
[R2-Serial1/0/0]quit
[R2]
```

2. 任务验证

(1) 主机连通性验证。

主机1、主机2和主机3能够相互连通，并能ping通公网服务器。其中主机2测试如图5-2所示。

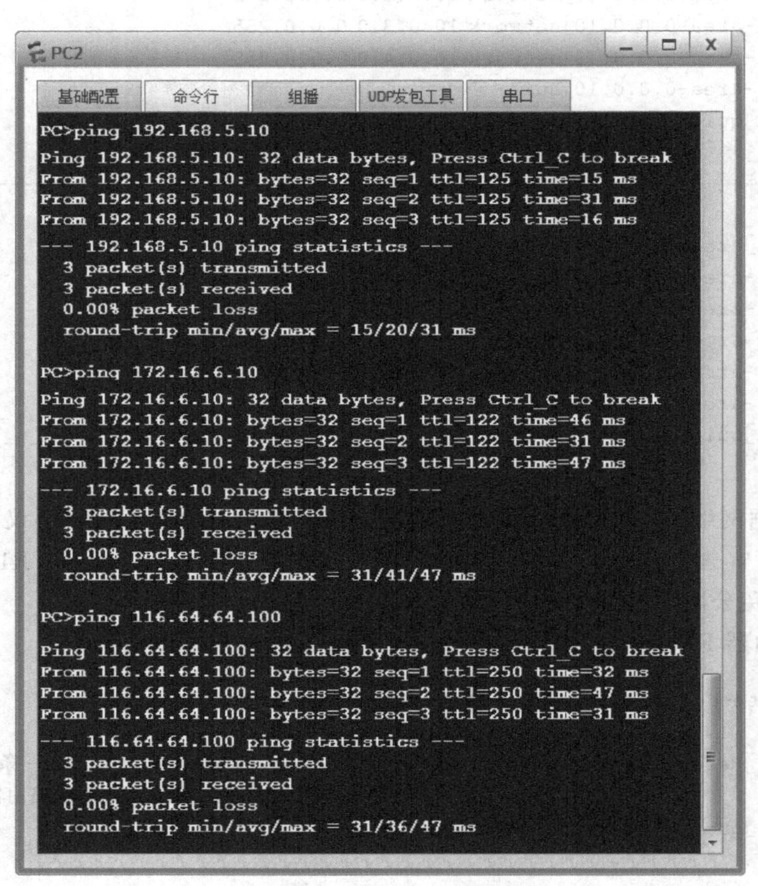

图5-2 主机2可以连通主机1、主机3和公网服务器

(2) 路由条目验证。

以路由器R7为例查看路由表。由于篇幅限制，路由表不列举直连路由条目。

```
[R7]display ip routing-table
Route Flags: R - relay, D - download to fib
------------------------------------------------------------------------------
Routing Tables: Public
         Destinations : 28        Routes : 29
Destination/Mask    Proto  Pre  Cost     Flags NextHop         Interface
        0.0.0.0/0   O_ASE  150  1          D   10.0.4.2        GigabitEthernet0/0/1
      10.0.1.0/24   OSPF   10   2          D   10.0.4.2        GigabitEthernet0/0/1
      10.0.2.0/24   OSPF   10   2          D   10.0.4.2        GigabitEthernet0/0/1
      10.0.3.0/24   OSPF   10   2          D   10.0.5.1        GigabitEthernet0/0/0
                    OSPF   10   2          D   10.0.4.2        GigabitEthernet0/0/1
    172.16.1.0/24   OSPF   10   4          D   10.0.4.2        GigabitEthernet0/0/1
    172.16.2.0/24   OSPF   10   4          D   10.0.4.2        GigabitEthernet0/0/1
    172.16.3.0/24   OSPF   10   5          D   10.0.4.2        GigabitEthernet0/0/1
    172.16.4.0/24   OSPF   10   5          D   10.0.4.2        GigabitEthernet0/0/1
    172.16.5.0/24   OSPF   10   6          D   10.0.4.2        GigabitEthernet0/0/1
    172.16.6.0/24   OSPF   10   6          D   10.0.4.2        GigabitEthernet0/0/1
   192.168.1.0/24   OSPF   10   3          D   10.0.4.2        GigabitEthernet0/0/1
   192.168.2.0/24   OSPF   10   4          D   10.0.4.2        GigabitEthernet0/0/1
   192.168.3.0/24   OSPF   10   3          D   10.0.4.2        GigabitEthernet0/0/1
   192.168.4.0/24   OSPF   10   3          D   10.0.4.2        GigabitEthernet0/0/1
   192.168.5.0/24   OSPF   10   3          D   10.0.4.2        GigabitEthernet0/0/1
```

可以看到 R7 有去往全网的 OSPF 路由条目。其中“0.0.0.0”是 OSPF 引入的外部默认路由，引入的 OSPF 外部路由协议用 O_ASE，优先级为 150。

【任务总结】

(1) 一个自治系统的 OSPF 区域最大路由上限为 30～200 个路由，一个 OSPF 路由表一般不允许装载超过 3 万条路由条目。

(2) 为避免区域间产生环路，常规区域必须与骨干区域连接，区域间路由条目由骨干区域边界路由器负责通告。

(3) 非骨干区域边界路由器不能通告区域间路由条目，只能向区域内部路由器或骨干边界路由器通告其学习到的 OSPF 路由条目。

工作任务六

OSPF 路由项过滤

【工作目的】

理解 IP 地址前缀和 ACL 的区别，掌握不同 OSPF 路由项过滤方式和配置过程。

【工作任务】

在上个工作任务中，A 公司收购 B 公司和 C 公司并划分出 3 个不同区域（公司 A 为 Area0，公司 B 为 Area10，公司 C 为 Area20），通过 OSPF 路由协议互联，外部通过公网 IP (202.116.64.1)连接至 Internet。为提高网络安全性，只允许总公司 A 能发现 B 公司和 C 公司网段信息，不允许 B 公司和 C 公司发现其他区域路由条目，但不影响其正常访问。具体任务如下。

(1) 在 R1、R3 过滤＜192.168.0.0＞和＜172.16.0.0＞路由项。

(2) 在 R2、R4 过滤＜192.168.0.0＞和＜10.0.0.0＞路由项。

(3) 在公司 A 骨干区域保存有所有区域路由条目，而公司 B 和公司 C 仅能看到本地区域路由条目。

(4) 路由过滤后不影响公司的正常访问，内网主机之间可以相互连通，并能通过公网 IP (202.116.64.1)访问公网服务器。

【工作背景】

OSPF 路由项过滤既可以用过滤策略 filter-policy 过滤，也可以由过滤器 filter 过滤。不同之处是过滤策略可以将所有区域指定路由条目一次过滤，简单方便；而过滤器可以在指定区域内精准过滤路由条目，如存在多个区域过滤配置的情况则显得较为烦琐。

【任务分析】

1. 地址前缀列表与过滤规则

地址前缀列表是一种包含一组路由信息过滤规则的过滤。ACL 和地址前缀列表都可以对路由进行匹配筛选。ACL 匹配路由时只能匹配路由的网络号，但无法匹配掩码，也就是前缀长度；而地址前缀列表比 ACL 更为灵活，可以匹配路由的网络号及掩码，当待过滤的路由已匹配当前表项的网络号时，掩码长度可以进行精确匹配或者在一定掩码长度范围内匹配。

地址前缀列表过滤路由的原则可以总结为：顺序匹配、唯一匹配、默认拒绝。

(1) 顺序匹配。按索引号从小到大顺序进行匹配。同一个地址前缀列表中的多条表项设置不同的索引号，可能会有不同的过滤结果，实际配置时需要注意。

(2) 唯一匹配。待过滤路由只要与一个表项匹配，就不会再去尝试匹配其他表项。

(3) 默认拒绝。默认所有未与任何一个表项匹配的路由都视为未通过地址前缀列表的过滤。因此在一个地址前缀列表中创建了一个或多个 deny 模式的表项后,需要创建一个表项来允许所有其他路由通过。

创建 IP 前缀列表语法如下,主要参数意义见表 6-1。

```
ip ip-prefix ip-prefix-name [index index-number] {permit|deny} ipv4-address mask-
length [greater-equal greater-equal-value] [less-equal less-equal-value]
```

表 6-1 主要参数意义

参 数	含 义
ipv4-address	用于指定网络号
mask-length	用于限定网络号的前多少位需严格匹配
greater-equal greater-equal-value	表示掩码大于或等于 greater-equal-value
less-equal less-equal-value	表示掩码小于或等于 less-equal-value

注:

① 如果不配置 greater-equal 和 less-equal 参数,则进行精确匹配,即只匹配掩码长度为 mask-length 的路由。

② 如果只配置 greater-equal 参数,则匹配的掩码长度范围为[greater-equal-value,32]。

③ 如果只配置 less-equal 参数,则匹配的掩码长度范围为[mask-length,less-equal-value]。

④ 如果同时配置 greater-equal 和 less-equal 参数,则匹配的掩码长度范围为[greater-equal-value,less-equal-value]。

2. OSPF 过滤方式

(1) 过滤策略过滤(注意:只能在 OSPF 进程中配置,不能在指定 area 区域中配置)。

① filter-policy import:该命令能抑制生成指定路由条目(不分区域,所有区域都过滤删除指定路由项,包括骨干区域 Area0),但不能抑制 LSA 泛洪。举例,如过滤和删除<172.16.0.0>路由项,脚本如下:

```
[Huawei]ip ip-prefix deny_172.16 deny  172.16.0.0 16 greater-equal 16
[Huawei]ip ip-prefix deny_172.16 permit 0.0.0.0 0 less-equal 32
[Huawei]ospf 1
[Huawei-ospf-1]filter-policy ip-prefix deny_172.16 import
[Huawei-ospf-1]quit
[Huawei]
```

② filter-policy export:该命令可以抑制指定 LSA 的泛洪,但 OSPF 的 LSA 一旦生成,该命令不起作用。

(2) 过滤器过滤(注意:只能在 OSPF 指定的 Area 区域中配置,不能在 OSPF 进程中配置)。

① filter import:该命令可以在具体 area 区域中抑制生成指定路由条目,但不能抑制区域间 LSA 泛洪。举例,如禁止 Area10 接收来自其他区域关于<172.16.0.0>网段 OSPF 通告(即删除 Area10 中<172.16.0.0>路由项),其脚本如下:

```
[Huawei]ip ip-prefix deny_172.16 deny  172.16.0.0 16 greater-equal 16
```

```
[Huawei]ip ip-prefix deny_172.16 permit 0.0.0.0 0 less-equal 32
[Huawei]ospf 1
[Huawei-ospf-1]area 10
[Huawei-ospf-1-area-0.0.0.10]
[R2-ospf-1-area-0.0.0.10]filter ip-prefix deny_172.16 import
[R2-ospf-1-area-0.0.0.10]quit
[R2-ospf-1]quit
[R2]
```

② filter export：该命令可以在具体 Area 区域中抑制指定 LSA 的泛洪。举例，如禁止骨干区域 Area0 向其他区域通告关于<172.16.0.0>网段的 LSA 分组(即删除其他所有区域的<172.16.0.0>路由项)。

```
[Huawei]ip ip-prefix deny_172.16 deny  172.16.0.0 16 greater-equal 16
[Huawei]ip ip-prefix deny_172.16 permit 0.0.0.0 0 less-equal 32
[Huawei]ospf 1
[Huawei-ospf-1]area 0
[R2-ospf-1-area-0.0.0.0]filter ip-prefix deny_172.16 export
[R2-ospf-1-area-0.0.0.0]quit
[R2-ospf-1]quit
[R2]
```

【设备器材】

路由器(AR1220)11 台，需要添加 1GEC 千兆接口模块或 2SA 串口模块。

主机 3 台，承担角色见表 6-2。

表 6-2　主机配置表

角　　色	接 入 方 式	IP 地址	所 属 公 司
主机 1	eNSP PC 接入	192.168.5.10/24	公司 A
主机 2	eNSP PC 接入	10.0.6.10/24	公司 B
主机 3	eNSP PC 接入	172.16.6.10/24	公司 C
公网服务器	eNSP Server 接入	116.64.64.100	

【环境拓扑】

环境拓扑如图 6-1 所示。

【工作过程】

1. 基本配置

(1) 路由器接口 IP 和系统名称配置。

请读者根据工作任务五配置多区域 OSPF 路由，实现 Area0、Area10 和 Area20 区域之间互通，并能通过公网 IP(202.116.64.1)访问公网服务器。

(2) 查看各区域 IP 路由表。

① 查看区域 Area10 路由项，以 R5 为例。

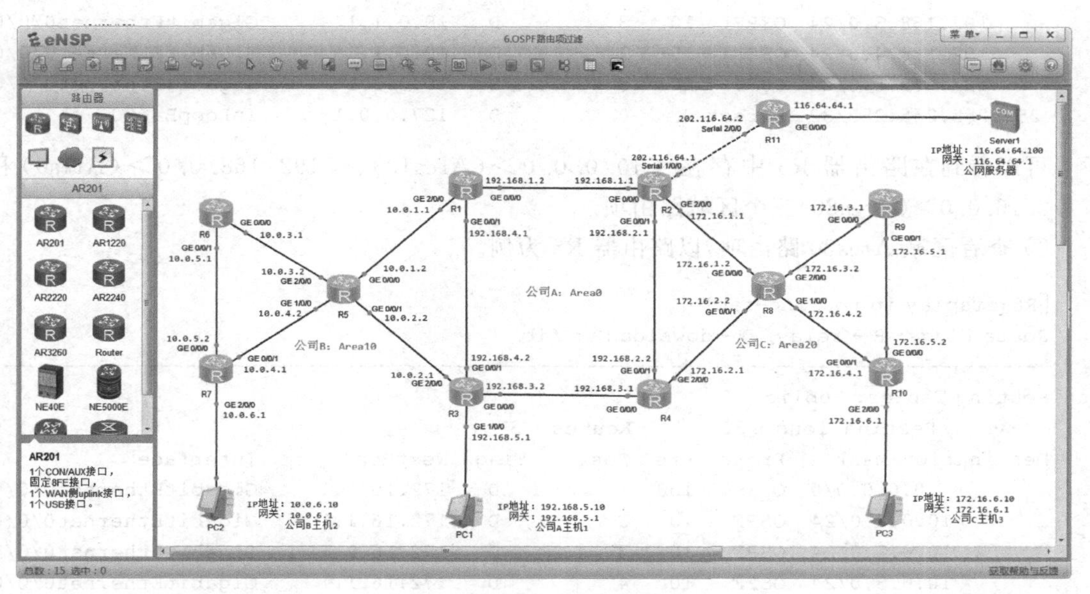

图 6-1 环境拓扑

```
[R5]display ip routing-table
Route Flags: R - relay, D - download to fib
------------------------------------------------------------------------------
Routing Tables: Public
         Destinations : 28       Routes : 29
Destination/Mask    Proto  Pre  Cost      Flags NextHop         Interface
          0.0.0.0/0 O_ASE  150  1           D   10.0.1.1        GigabitEthernet0/0/0
        10.0.1.0/24 Direct 0    0           D   10.0.1.2        GigabitEthernet0/0/0
        10.0.1.2/32 Direct 0    0           D   127.0.0.1       GigabitEthernet0/0/0
      10.0.1.255/32 Direct 0    0           D   127.0.0.1       GigabitEthernet0/0/0
        10.0.2.0/24 Direct 0    0           D   10.0.2.2        GigabitEthernet0/0/1
        10.0.2.2/32 Direct 0    0           D   127.0.0.1       GigabitEthernet0/0/1
      10.0.2.255/32 Direct 0    0           D   127.0.0.1       GigabitEthernet0/0/1
        10.0.3.0/24 Direct 0    0           D   10.0.3.2        GigabitEthernet2/0/0
        10.0.3.2/32 Direct 0    0           D   127.0.0.1       GigabitEthernet2/0/0
      10.0.3.255/32 Direct 0    0           D   127.0.0.1       GigabitEthernet2/0/0
        10.0.4.0/24 Direct 0    0           D   10.0.4.2        GigabitEthernet1/0/0
        10.0.4.2/32 Direct 0    0           D   127.0.0.1       GigabitEthernet1/0/0
      10.0.4.255/32 Direct 0    0           D   127.0.0.1       GigabitEthernet1/0/0
        10.0.5.0/24 OSPF   10   2           D   10.0.3.1        GigabitEthernet2/0/0
                    OSPF   10   2           D   10.0.4.1        GigabitEthernet1/0/0
        10.0.6.0/24 OSPF   10   2           D   10.0.4.1        GigabitEthernet1/0/0
        127.0.0.0/8 Direct 0    0           D   127.0.0.1       InLoopBack0
       127.0.0.1/32 Direct 0    0           D   127.0.0.1       InLoopBack0
 127.255.255.255/32 Direct 0    0           D   127.0.0.1       InLoopBack0
      172.16.1.0/24 OSPF   10   3           D   10.0.1.1        GigabitEthernet0/0/0
      172.16.2.0/24 OSPF   10   4           D   10.0.1.1        GigabitEthernet0/0/0
      172.16.3.0/24 OSPF   10   4           D   10.0.1.1        GigabitEthernet0/0/0
      172.16.4.0/24 OSPF   10   4           D   10.0.1.1        GigabitEthernet0/0/0
     192.168.1.0/24 OSPF   10   2           D   10.0.1.1        GigabitEthernet0/0/0
     192.168.2.0/24 OSPF   10   3           D   10.0.1.1        GigabitEthernet0/0/0
```

```
      192.168.3.0/24  OSPF    10   3           D   10.0.1.1        GigabitEthernet0/0/0
      192.168.4.0/24  OSPF    10   2           D   10.0.1.1        GigabitEthernet0/0/0
      192.168.5.0/24  OSPF    10   3           D   10.0.1.1        GigabitEthernet0/0/0
  255.255.255.255/32  Direct  0    0           D   127.0.0.1       InLoopBack0
```

可以看到在路由器 R5 中存在<10. 0. 0. 0>(Area10)、<192. 168. 0. 0>(Area0)和<172. 16. 0. 0>(Area20)三个区域路由项。

② 查看区域 Area20 路由项,以路由器 R8 为例。

```
[R8]display ip routing-table
Route Flags: R - relay, D - download to fib
------------------------------------------------------------------------------
Routing Tables: Public
         Destinations : 30        Routes : 37
Destination/Mask    Proto   Pre  Cost      Flags NextHop         Interface
         0.0.0.0/0  O_ASE   150  1           D   172.16.1.1      GigabitEthernet0/0/0
       10.0.1.0/24  OSPF    10   3           D   172.16.1.1      GigabitEthernet0/0/0
       10.0.2.0/24  OSPF    10   3           D   172.16.2.1      GigabitEthernet0/0/1
       10.0.3.0/24  OSPF    10   4           D   172.16.1.1      GigabitEthernet0/0/0
                    OSPF    10   4           D   172.16.2.1      GigabitEthernet0/0/1
       10.0.4.0/24  OSPF    10   4           D   172.16.1.1      GigabitEthernet0/0/0
                    OSPF    10   4           D   172.16.2.1      GigabitEthernet0/0/1
       10.0.5.0/24  OSPF    10   5           D   172.16.1.1      GigabitEthernet0/0/0
                    OSPF    10   5           D   172.16.2.1      GigabitEthernet0/0/1
       10.0.6.0/24  OSPF    10   5           D   172.16.1.1      GigabitEthernet0/0/0
                    OSPF    10   5           D   172.16.2.1      GigabitEthernet0/0/1
       127.0.0.0/8  Direct  0    0           D   127.0.0.1       InLoopBack0
      127.0.0.1/32  Direct  0    0           D   127.0.0.1       InLoopBack0
127.255.255.255/32  Direct  0    0           D   127.0.0.1       InLoopBack0
     172.16.1.0/24  Direct  0    0           D   172.16.1.2      GigabitEthernet0/0/0
     172.16.1.2/32  Direct  0    0           D   127.0.0.1       GigabitEthernet0/0/0
   172.16.1.255/32  Direct  0    0           D   127.0.0.1       GigabitEthernet0/0/0
     172.16.2.0/24  Direct  0    0           D   172.16.2.2      GigabitEthernet0/0/1
     172.16.2.2/32  Direct  0    0           D   127.0.0.1       GigabitEthernet0/0/1
   172.16.2.255/32  Direct  0    0           D   127.0.0.1       GigabitEthernet0/0/1
     172.16.3.0/24  Direct  0    0           D   172.16.3.2      GigabitEthernet2/0/0
     172.16.3.2/32  Direct  0    0           D   127.0.0.1       GigabitEthernet2/0/0
   172.16.3.255/32  Direct  0    0           D   127.0.0.1       GigabitEthernet2/0/0
     172.16.4.0/24  Direct  0    0           D   172.16.4.2      GigabitEthernet1/0/0
     172.16.4.2/32  Direct  0    0           D   127.0.0.1       GigabitEthernet1/0/0
   172.16.4.255/32  Direct  0    0           D   127.0.0.1       GigabitEthernet1/0/0
     172.16.5.0/24  OSPF    10   2           D   172.16.3.1      GigabitEthernet2/0/0
                    OSPF    10   2           D   172.16.4.1      GigabitEthernet1/0/0
     172.16.6.0/24  OSPF    10   2           D   172.16.4.1      GigabitEthernet1/0/0
    192.168.1.0/24  OSPF    10   2           D   172.16.1.1      GigabitEthernet0/0/0
    192.168.2.0/24  OSPF    10   2           D   172.16.2.1      GigabitEthernet0/0/1
                    OSPF    10   2           D   172.16.1.1      GigabitEthernet0/0/0
    192.168.3.0/24  OSPF    10   2           D   172.16.2.1      GigabitEthernet0/0/1
    192.168.4.0/24  OSPF    10   3           D   172.16.1.1      GigabitEthernet0/0/0
                    OSPF    10   3           D   172.16.2.1      GigabitEthernet0/0/1
    192.168.5.0/24  OSPF    10   3           D   172.16.2.1      GigabitEthernet0/0/1
255.255.255.255/32  Direct  0    0           D   127.0.0.1       InLoopBack0
```

在路由器 R8 中同样存在<10.0.0.0>(Area10)、<192.168.0.0>(Area0)和<172.16.0.0>(Area20)三个区域路由项。

③ 查看骨干区域 Area0 路由项，以 R3 为例。

```
[R3]display ip routing-table
Route Flags: R - relay, D - download to fib
------------------------------------------------------------------------------
Routing Tables: Public
         Destinations : 30        Routes : 32
Destination/Mask    Proto  Pre  Cost     Flags NextHop         Interface
        0.0.0.0/0   O_ASE  150  1          D   192.168.4.1     GigabitEthernet0/0/1
                    O_ASE  150  1          D   192.168.3.1     GigabitEthernet0/0/0
      10.0.1.0/24   OSPF   10   2          D   10.0.2.2        GigabitEthernet2/0/0
      10.0.2.0/24   Direct 0    0          D   10.0.2.1        GigabitEthernet2/0/0
      10.0.2.1/32   Direct 0    0          D   127.0.0.1       GigabitEthernet2/0/0
    10.0.2.255/32   Direct 0    0          D   127.0.0.1       GigabitEthernet2/0/0
      10.0.3.0/24   OSPF   10   2          D   10.0.2.2        GigabitEthernet2/0/0
      10.0.4.0/24   OSPF   10   2          D   10.0.2.2        GigabitEthernet2/0/0
      10.0.5.0/24   OSPF   10   3          D   10.0.2.2        GigabitEthernet2/0/0
      10.0.6.0/24   OSPF   10   3          D   10.0.2.2        GigabitEthernet2/0/0
      127.0.0.0/8   Direct 0    0          D   127.0.0.1       InLoopBack0
     127.0.0.1/32   Direct 0    0          D   127.0.0.1       InLoopBack0
127.255.255.255/32  Direct 0    0          D   127.0.0.1       InLoopBack0
    172.16.1.0/24   OSPF   10   3          D   192.168.4.1     GigabitEthernet0/0/1
                    OSPF   10   3          D   192.168.3.1     GigabitEthernet0/0/0
    172.16.2.0/24   OSPF   10   2          D   192.168.3.1     GigabitEthernet0/0/0
    172.16.3.0/24   OSPF   10   3          D   192.168.3.1     GigabitEthernet0/0/0
    172.16.4.0/24   OSPF   10   3          D   192.168.3.1     GigabitEthernet0/0/0
    172.16.5.0/24   OSPF   10   4          D   192.168.3.1     GigabitEthernet0/0/0
    172.16.6.0/24   OSPF   10   4          D   192.168.3.1     GigabitEthernet0/0/0
   192.168.1.0/24   OSPF   10   2          D   192.168.4.1     GigabitEthernet0/0/1
   192.168.2.0/24   OSPF   10   2          D   192.168.3.1     GigabitEthernet0/0/0
   192.168.3.0/24   Direct 0    0          D   192.168.3.2     GigabitEthernet0/0/0
   192.168.3.2/32   Direct 0    0          D   127.0.0.1       GigabitEthernet0/0/0
 192.168.3.255/32   Direct 0    0          D   127.0.0.1       GigabitEthernet0/0/0
   192.168.4.0/24   Direct 0    0          D   192.168.4.2     GigabitEthernet0/0/1
   192.168.4.2/32   Direct 0    0          D   127.0.0.1       GigabitEthernet0/0/1
 192.168.4.255/32   Direct 0    0          D   127.0.0.1       GigabitEthernet0/0/1
   192.168.5.0/24   Direct 0    0          D   192.168.5.1     GigabitEthernet1/0/0
   192.168.5.1/32   Direct 0    0          D   127.0.0.1       GigabitEthernet1/0/0
 192.168.5.255/32   Direct 0    0          D   127.0.0.1       GigabitEthernet1/0/0
255.255.255.255/32  Direct 0    0          D   127.0.0.1       InLoopBack0
```

路由器 R3 同样存在<10.0.0.0>(Area10)、<192.168.0.0>(Area0)和<172.16.0.0>(Area20)三个区域路由项。

(3) 配置 OSPF 路由项过滤(请读者选择以下任意一种方法)。

方法是在 Area10 和 Area20 区域中的 OSPF LSA 入栈方向过滤，禁止 Area10 和 Area20 收到来自骨干区域 Area0 中含有<192.168.0.0>、<172.16.0.0>或<10.0.0.0>OSPF 路由项。

① 在公司 A 路由器 R1 和 R3 过滤<192.168.0.0>(Area0)和<172.16.0.0>(Area20)

路由项。

```
[R1]ip ip-prefix deny_192_172 index 10 deny 192.168.0.0 16 greater-equal 16
[R1]ip ip-prefix deny_192_172 index 20 deny 172.16.0.0 16 greater-equal 16
[R1]ip ip-prefix deny_192_172 index 30 permit 0.0.0.0 0 less-equal 32
[R1]ospf 1
[R1-ospf-1]area 10
[R1-ospf-1-area-0.0.0.10]filter ip-prefix deny_192_172 import   //Area10 从 Area0 接
                                                                  收路由通告时过滤
[R1-ospf-1-area-0.0.0.10]quit
[R1-ospf-1]quit
[R1]
-------------------------------------------------------------------------------
[R3]ip ip-prefix deny_192_172 index 10 deny 192.168.0.0 16 greater-equal 16
[R3]ip ip-prefix deny_192_172 index 20 deny 172.16.0.0 16 greater-equal 16
[R3]ip ip-prefix deny_192_172 index 30 permit 0.0.0.0 0 less-equal 32
[R3]ospf 1
[R3-ospf-1]area 10
[R3-ospf-1-area-0.0.0.10]filter ip-prefix deny_192_172 import
[R3-ospf-1-area-0.0.0.10]quit
[R3-ospf-1]quit
[R3]
```

② 在公司 A 路由器 R2 和 R4 过滤<192.168.0.0>(Area0)和<10.0.0.0>(Area10)路由项。

```
[R2]ip ip-prefix deny_192_10 deny 192.168.0.0 16 greater-equal 16
[R2]ip ip-prefix deny_192_10 deny 10.0.0.0 8 greater-equal 8
[R2]ip ip-prefix deny_192_10 permit 0.0.0.0 0 less-equal 32
[R2]ospf 1
[R2-ospf-1]area 20
[R2-ospf-1-area-0.0.0.20]filter ip-prefix deny_192_10 import   //Area20 从 Area0 接收
                                                                路由通告时过滤
[R2-ospf-1-area-0.0.0.20]quit
[R2-ospf-1]quit
[R2]
-------------------------------------------------------------------------------
[R4]ip ip-prefix deny_192_10 deny 192.168.0.0 16 greater-equal 16
[R4]ip ip-prefix deny_192_10 deny 10.0.0.0 8 greater-equal 8
[R4]ip ip-prefix deny_192_10 permit 0.0.0.0 0 less-equal 32
[R4]ospf 1
[R4-ospf-1]area 20
[R4-ospf-1-area-0.0.0.20]filter ip-prefix deny_192_10 import
[R4-ospf-1-area-0.0.0.20]quit
[R4-ospf-1]quit
[R4]
```

也可以在骨干区域 Area0 的 OSPF LSA 出栈方向过滤<192.168.0.0>、<172.16.0.0>或<10.0.0.0>OSPF 路由项，从而禁止 Area10 和 Area20 接收这些路由项。

```
[R1]ip ip-prefix deny_192_172 index 10 deny 192.168.0.0 16 greater-equal 16
[R1]ip ip-prefix deny_192_172 index 20 deny 172.16.0.0 16 greater-equal 16
[R1]ip ip-prefix deny_192_172 index 30 permit 0.0.0.0 0 less-equal 32
```

```
[R1]ospf 1
[R1-ospf-1]area 0
[R1-ospf-1-area-0.0.0.0]filter ip-prefix deny_192_172 export  //Area0 向其他 Area 通
                                                                告时过滤
[R1-ospf-1-area-0.0.0.0]quit
[R1-ospf-1]quit
[R1]
--------------------------------------------------------------------------------
[R3]ip ip-prefix deny_192_172 index 10 deny 192.168.0.0 16 greater-equal 16
[R3]ip ip-prefix deny_192_172 index 20 deny 172.16.0.0 16 greater-equal 16
[R3]ip ip-prefix deny_192_172 index 30 permit 0.0.0.0 0 less-equal 32
[R3]ospf 1
[R3-ospf-1]area 0
[R3-ospf-1-area-0.0.0.0]filter ip-prefix deny_192_172 export  //Area0 向其他 Area 通
                                                                告时过滤

[R3-ospf-1-area-0.0.0.0]quit
[R3-ospf-1]quit
[R3]
--------------------------------------------------------------------------------
[R2]ip ip-prefix deny_192_10 deny 192.168.0.0 16 greater-equal 16
[R2]ip ip-prefix deny_192_10 deny 10.0.0.0 8 greater-equal 8
[R2]ip ip-prefix deny_192_10 permit 0.0.0.0 0 less-equal 32
[R2]ospf 1
[R2-ospf-1]area 0
[R2-ospf-1-area-0.0.0.0]filter ip-prefix deny_192_10 export  //Area0 向其他 Area 通
                                                               告时过滤

[R2-ospf-1-area-0.0.0.0]quit
[R2-ospf-1]quit
[R2]
--------------------------------------------------------------------------------
[R4]ip ip-prefix deny_192_10 deny 192.168.0.0 16 greater-equal 16
[R4]ip ip-prefix deny_192_10 deny 10.0.0.0 8 greater-equal 8
[R4]ip ip-prefix deny_192_10 permit 0.0.0.0 0 less-equal 32
[R4]ospf 1
[R4-ospf-1]area 0
[R4-ospf-1-area-0.0.0.0]filter ip-prefix deny_192_10 export  //Area0 向其他 Area 通
                                                               告时过滤

[R4-ospf-1-area-0.0.0.0]quit
[R4-ospf-1]quit
[R4]
```

2. 任务验证

(1) 路由条目验证。

① 验证区域 10 路由项，以 R5 为例。

```
[R5]display ip routing-table
Route Flags: R - relay, D - download to fib
--------------------------------------------------------------------------------
```

```
Routing Tables: Public
         Destinations : 19        Routes : 20
  Destination/Mask    Proto  Pre  Cost     Flags NextHop         Interface
        0.0.0.0/0   O_ASE  150  1           D   10.0.1.1        GigabitEthernet0/0/0
      10.0.1.0/24   Direct 0    0           D   10.0.1.2        GigabitEthernet0/0/0
      10.0.1.2/32   Direct 0    0           D   127.0.0.1       GigabitEthernet0/0/0
    10.0.1.255/32   Direct 0    0           D   127.0.0.1       GigabitEthernet0/0/0
      10.0.2.0/24   Direct 0    0           D   10.0.2.2        GigabitEthernet0/0/1
      10.0.2.2/32   Direct 0    0           D   127.0.0.1       GigabitEthernet0/0/1
    10.0.2.255/32   Direct 0    0           D   127.0.0.1       GigabitEthernet0/0/1
      10.0.3.0/24   Direct 0    0           D   10.0.3.2        GigabitEthernet2/0/0
      10.0.3.2/32   Direct 0    0           D   127.0.0.1       GigabitEthernet2/0/0
    10.0.3.255/32   Direct 0    0           D   127.0.0.1       GigabitEthernet2/0/0
      10.0.4.0/24   Direct 0    0           D   10.0.4.2        GigabitEthernet1/0/0
      10.0.4.2/32   Direct 0    0           D   127.0.0.1       GigabitEthernet1/0/0
    10.0.4.255/32   Direct 0    0           D   127.0.0.1       GigabitEthernet1/0/0
      10.0.5.0/24   OSPF   10   2           D   10.0.3.1        GigabitEthernet2/0/0
                    OSPF   10   2           D   10.0.4.1        GigabitEthernet1/0/0
      10.0.6.0/24   OSPF   10   2           D   10.0.4.1        GigabitEthernet1/0/0
      127.0.0.0/8   Direct 0    0           D   127.0.0.1       InLoopBack0
     127.0.0.1/32   Direct 0    0           D   127.0.0.1       InLoopBack0
127.255.255.255/32  Direct 0    0           D   127.0.0.1       InLoopBack0
255.255.255.255/32  Direct 0    0           D   127.0.0.1       InLoopBack0
```

可以看到，路由器 R5 仅存在本区域路由，<192.168.0.0>（Area0）和<172.16.0.0>（Area20）路由项已全部过滤。其中，0.0.0.0 是 OSPF 引入的外部默认路由未被过滤。

② 验证区域 20 路由项，以 R8 为例。

```
[R8]display ip routing-table
Route Flags: R - relay, D - download to fib
------------------------------------------------------------------------------
Routing Tables: Public
         Destinations : 19        Routes : 20
  Destination/Mask    Proto  Pre  Cost     Flags NextHop         Interface
        0.0.0.0/0   O_ASE  150  1           D   172.16.1.1      GigabitEthernet0/0/0
      127.0.0.0/8   Direct 0    0           D   127.0.0.1       InLoopBack0
     127.0.0.1/32   Direct 0    0           D   127.0.0.1       InLoopBack0
127.255.255.255/32  Direct 0    0           D   127.0.0.1       InLoopBack0
    172.16.1.0/24   Direct 0    0           D   172.16.1.2      GigabitEthernet0/0/0
    172.16.1.2/32   Direct 0    0           D   127.0.0.1       GigabitEthernet0/0/0
  172.16.1.255/32   Direct 0    0           D   127.0.0.1       GigabitEthernet0/0/0
    172.16.2.0/24   Direct 0    0           D   172.16.2.2      GigabitEthernet0/0/1
    172.16.2.2/32   Direct 0    0           D   127.0.0.1       GigabitEthernet0/0/1
  172.16.2.255/32   Direct 0    0           D   127.0.0.1       GigabitEthernet0/0/1
    172.16.3.0/24   Direct 0    0           D   172.16.3.2      GigabitEthernet2/0/0
    172.16.3.2/32   Direct 0    0           D   127.0.0.1       GigabitEthernet2/0/0
  172.16.3.255/32   Direct 0    0           D   127.0.0.1       GigabitEthernet2/0/0
    172.16.4.0/24   Direct 0    0           D   172.16.4.2      GigabitEthernet1/0/0
    172.16.4.2/32   Direct 0    0           D   127.0.0.1       GigabitEthernet1/0/0
  172.16.4.255/32   Direct 0    0           D   127.0.0.1       GigabitEthernet1/0/0
    172.16.5.0/24   OSPF   10   2           D   172.16.3.1      GigabitEthernet2/0/0
                    OSPF   10   2           D   172.16.4.1      GigabitEthernet1/0/0
```

```
       172.16.6.0/24  OSPF    10   2          D   172.16.4.1      GigabitEthernet1/0/0
  255.255.255.255/32  Direct  0    0          D   127.0.0.1       InLoopBack0
```

路由器 R8 仅存在本区域路由，<192.168.0.0>(Area0)和<10.0.0.0>(Area10)路由项已全部过滤。其中，0.0.0.0 是 OSPF 引入的外部默认路由未被过滤。

③ 验证骨干区域 Area0 路由项，以 R3 为例。

```
[R3]display ip routing-table
Route Flags: R - relay, D - download to fib
------------------------------------------------------------------------------
Routing Tables: Public
         Destinations : 30       Routes : 32
  Destination/Mask    Proto   Pre  Cost      Flags NextHop         Interface
         0.0.0.0/0    O_ASE   150  1          D   192.168.3.1     GigabitEthernet0/0/0
                      O_ASE   150  1          D   192.168.4.1     GigabitEthernet0/0/1
        10.0.1.0/24   OSPF    10   2          D   10.0.2.2        GigabitEthernet2/0/0
        10.0.2.0/24   Direct  0    0          D   10.0.2.1        GigabitEthernet2/0/0
        10.0.2.1/32   Direct  0    0          D   127.0.0.1       GigabitEthernet2/0/0
      10.0.2.255/32   Direct  0    0          D   127.0.0.1       GigabitEthernet2/0/0
        10.0.3.0/24   OSPF    10   2          D   10.0.2.2        GigabitEthernet2/0/0
        10.0.4.0/24   OSPF    10   2          D   10.0.2.2        GigabitEthernet2/0/0
        10.0.5.0/24   OSPF    10   3          D   10.0.2.2        GigabitEthernet2/0/0
        10.0.6.0/24   OSPF    10   3          D   10.0.2.2        GigabitEthernet2/0/0
        127.0.0.0/8   Direct  0    0          D   127.0.0.1       InLoopBack0
       127.0.0.1/32   Direct  0    0          D   127.0.0.1       InLoopBack0
 127.255.255.255/32   Direct  0    0          D   127.0.0.1       InLoopBack0
      172.16.1.0/24   OSPF    10   3          D   192.168.3.1     GigabitEthernet0/0/0
                      OSPF    10   3          D   192.168.4.1     GigabitEthernet0/0/1
      172.16.2.0/24   OSPF    10   2          D   192.168.3.1     GigabitEthernet0/0/0
      172.16.3.0/24   OSPF    10   3          D   192.168.3.1     GigabitEthernet0/0/0
      172.16.4.0/24   OSPF    10   3          D   192.168.3.1     GigabitEthernet0/0/0
      172.16.5.0/24   OSPF    10   4          D   192.168.3.1     GigabitEthernet0/0/0
      172.16.6.0/24   OSPF    10   4          D   192.168.3.1     GigabitEthernet0/0/0
     192.168.1.0/24   OSPF    10   2          D   192.168.4.1     GigabitEthernet0/0/1
     192.168.2.0/24   OSPF    10   2          D   192.168.3.1     GigabitEthernet0/0/0
     192.168.3.0/24   Direct  0    0          D   192.168.3.2     GigabitEthernet0/0/0
     192.168.3.2/32   Direct  0    0          D   127.0.0.1       GigabitEthernet0/0/0
   192.168.3.255/32   Direct  0    0          D   127.0.0.1       GigabitEthernet0/0/0
     192.168.4.0/24   Direct  0    0          D   192.168.4.2     GigabitEthernet0/0/1
     192.168.4.2/32   Direct  0    0          D   127.0.0.1       GigabitEthernet0/0/1
   192.168.4.255/32   Direct  0    0          D   127.0.0.1       GigabitEthernet0/0/1
     192.168.5.0/24   Direct  0    0          D   192.168.5.1     GigabitEthernet1/0/0
     192.168.5.1/32   Direct  0    0          D   127.0.0.1       GigabitEthernet1/0/0
   192.168.5.255/32   Direct  0    0          D   127.0.0.1       GigabitEthernet1/0/0
 255.255.255.255/32   Direct  0    0          D   127.0.0.1       InLoopBack0
```

路由器 R3 路由条目不受影响，仍然存在<10.0.0.0>(Area10)、<192.168.0.0>(Area0)和<172.16.0.0>(Area20)三个区域路由项。

(2) 连通性验证。

主机 1、主机 2 和主机 3 能够相互连通，并能 ping 通公网服务器。其中，主机 2 测试如图 6-2 所示。

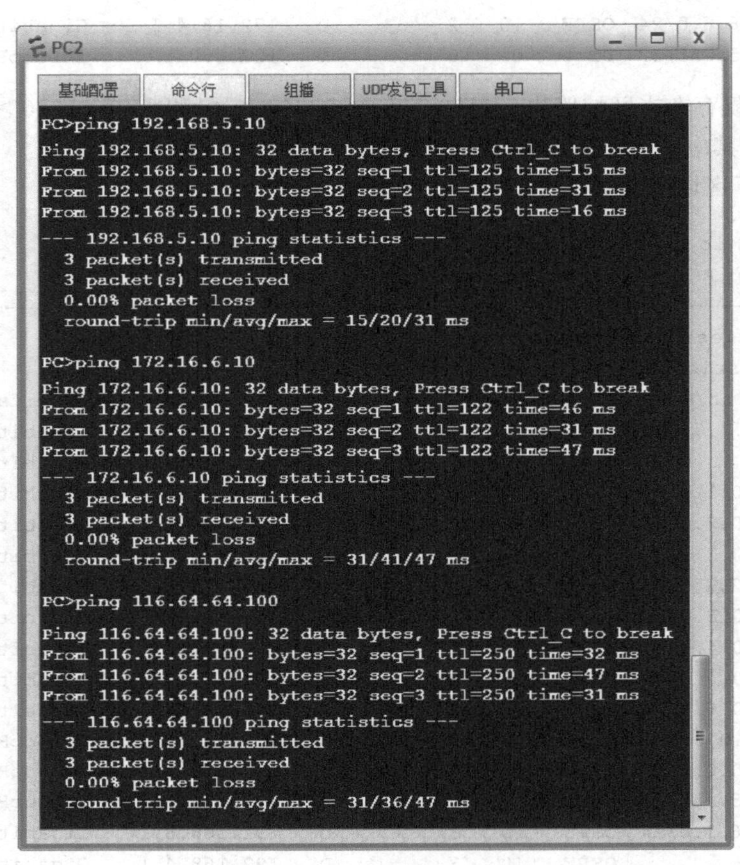

图 6-2 主机 2 可以连通主机 1、主机 3 和公网服务器

【任务总结】

(1) IP 地址前缀比 ACL 更为灵活,可以匹配一段掩码。假如匹配掩码固定,本例也可以用 ACL 实现,脚本如下:

```
[R1]acl name deny_192_172          //后面不加 advance 或 basic 参数,默认 advance
[R1-acl-adv-deny_192_172]rule 10 deny ip source 192.168.0.0 0.0.255.255
[R1-acl-adv-deny_192_172]rule 20 deny ip source 172.16.0.0 0.0.255.255
[R1-acl-adv-deny_192_172]rule 30 permit ip source 0.0.0.0 0.0.0.0
[R1-acl-adv-deny_192_172]quit
[R1]ospf 1
[R1-ospf-1]area 10
[R1-ospf-1-area-0.0.0.10]filter acl-name deny_192_172 import
[R1-ospf-1-area-0.0.0.10]quit
[R1-ospf-1]quit
[R1]
------------------------------------------------------------------------------------
[R3]acl name deny_192_172
[R3-acl-adv-gdcp]rule 10 deny ip source 192.168.0.0 0.0.255.255
[R3-acl-adv-gdcp]rule 20 deny ip source 172.16.0.0 0.0.255.255
[R3-acl-adv-gdcp]rule 30 permit ip source 0.0.0.0 0.0.0.0
```

```
[R3-acl-adv-gdcp]quit
[R3]ospf 1
[R3-ospf-1]area 10
[R3-ospf-1-area-0.0.0.10]filter acl-name deny_192_172 import
[R3-ospf-1-area-0.0.0.10]quit
[R3-ospf-1]quit
[R3]
--------------------------------------------------------------------------------
[R2]acl name deny_192_10 basic
[R2-acl-basic-deny_192_10]rule 10 deny source 192.168.0.0 0.0.255.255
[R2-acl-basic-deny_192_10]rule 20 deny source 10.0.0.0 0.255.255.255
[R2-acl-basic-deny_192_10]rule 30 permit source 0.0.0.0 0.0.0.0
[R2-acl-basic-deny_192_10]quit
[R2]ospf 1
[R2-ospf-1]area 20
[R2-ospf-1-area-0.0.0.20]filter acl-name deny_192_10 import
[R2-ospf-1-area-0.0.0.20]quit
[R2-ospf-1]quit
[R2]
--------------------------------------------------------------------------------
[R4]acl name deny_192_10 basic
[R4-acl-basic-deny_192_10]rule 10 deny source 192.168.0.0 0.0.255.255
[R4-acl-basic-deny_192_10]rule 20 deny source 10.0.0.0 0.255.255.255
[R4-acl-basic-deny_192_10]rule 30 permit source 0.0.0.0 0.0.0.0
[R4-acl-basic-deny_192_10]quit
[R4]ospf 1
[R4-ospf-1]area 20
[R4-ospf-1-area-0.0.0.20]filter acl-name deny_192_10 import
[R4-ospf-1-area-0.0.0.20]quit
[R4-ospf-1]quit
[R4]
```

路由表与连通性测试结果与用 IP 地址前缀实验一致。

(2) 利用过滤策略可以将所有区域指定路由条目一次过滤，而过滤器可在指定区域内精准过滤路由条目，应根据实际场合选择适当过滤方式。在本工作任务中，假如使用过滤策略过滤 OSPF 路由项，会出现部分区域无法连通问题。

以下为错误配置脚本。

① 在公司 A 路由器 R1 和 R3 过滤<192.168.0.0>(Area0)和<172.16.0.0>(Area20)路由项。

```
[R1]ip ip-prefix deny_192_172 index 10 deny 192.168.0.0 16 greater-equal 16
[R1]ip ip-prefix deny_192_172 index 20 deny 172.16.0.0 16 greater-equal 16
[R1]ip ip-prefix deny_192_172 index 30 permit 0.0.0.0 0 less-equal 32
[R1]ospf 1
[R1-ospf-1]filter-policy ip-prefix deny_192_172 import
[R1-ospf-1]quit
[R1]
--------------------------------------------------------------------------------
[R3]ip ip-prefix deny_192_172 index 10 deny 192.168.0.0 16 greater-equal 16
[R3]ip ip-prefix deny_192_172 index 20 deny 172.16.0.0 16 greater-equal 16
[R3]ip ip-prefix deny_192_172 index 30 permit 0.0.0.0 0 less-equal 32
```

```
[R3]ospf 1
[R3-ospf-1]filter-policy ip-prefix deny_192_172 import
[R3-ospf-1]quit
[R3]
```

② 在公司 A 路由器 R2 和 R4 过滤<192.168.0.0>(Area0)和<10.0.0.0>(Area10)路由项。

```
[R2]ip ip-prefix deny_192_10 deny 192.168.0.0 16 greater-equal 16
[R2]ip ip-prefix deny_192_10 deny 10.0.0.0 8 greater-equal 8
[R2]ip ip-prefix deny_192_10 permit 0.0.0.0 0 less-equal 32
[R2]ospf 1
[R2-ospf-1]filter-policy ip-prefix deny_192_10 import
[R2-ospf-1]quit
[R2]
--------------------------------------------------------------------------------
[R4]ip ip-prefix deny_192_10 deny 192.168.0.0 16 greater-equal 16
[R4]ip ip-prefix deny_192_10 deny 10.0.0.0 8 greater-equal 8
[R4]ip ip-prefix deny_192_10 permit 0.0.0.0 0 less-equal 32
[R4]ospf 1
[R4-ospf-1]filter-policy ip-prefix deny_192_10 import
[R4-ospf-1]quit
[R4]
```

③ 路由条目验证。

- 验证区域 10 路由项,以 R5 为例。

```
[R5]display ip routing-table
Route Flags: R - relay, D - download to fib
--------------------------------------------------------------------------------
Routing Tables: Public
         Destinations : 19        Routes : 20
  Destination/Mask    Proto   Pre  Cost      Flags NextHop         Interface
          0.0.0.0/0   O_ASE   150  1           D   10.0.1.1        GigabitEthernet0/0/0
        10.0.1.0/24   Direct  0    0           D   10.0.1.2        GigabitEthernet0/0/0
        10.0.1.2/32   Direct  0    0           D   127.0.0.1       GigabitEthernet0/0/0
      10.0.1.255/32   Direct  0    0           D   127.0.0.1       GigabitEthernet0/0/0
        10.0.2.0/24   Direct  0    0           D   10.0.2.2        GigabitEthernet0/0/1
        10.0.2.2/32   Direct  0    0           D   127.0.0.1       GigabitEthernet0/0/1
      10.0.2.255/32   Direct  0    0           D   127.0.0.1       GigabitEthernet0/0/1
        10.0.3.0/24   Direct  0    0           D   10.0.3.2        GigabitEthernet2/0/0
        10.0.3.2/32   Direct  0    0           D   127.0.0.1       GigabitEthernet2/0/0
      10.0.3.255/32   Direct  0    0           D   127.0.0.1       GigabitEthernet2/0/0
        10.0.4.0/24   Direct  0    0           D   10.0.4.2        GigabitEthernet1/0/0
        10.0.4.2/32   Direct  0    0           D   127.0.0.1       GigabitEthernet1/0/0
      10.0.4.255/32   Direct  0    0           D   127.0.0.1       GigabitEthernet1/0/0
        10.0.5.0/24   OSPF    10   2           D   10.0.3.1        GigabitEthernet2/0/0
                      OSPF    10   2           D   10.0.4.1        GigabitEthernet1/0/0
        10.0.6.0/24   OSPF    10   2           D   10.0.4.1        GigabitEthernet1/0/0
        127.0.0.0/8   Direct  0    0           D   127.0.0.1       InLoopBack0
       127.0.0.1/32   Direct  0    0           D   127.0.0.1       InLoopBack0
 127.255.255.255/32   Direct  0    0           D   127.0.0.1       InLoopBack0
 255.255.255.255/32   Direct  0    0           D   127.0.0.1       InLoopBack0
```

可以看到，在路由器 R5 上仅存有本区域路由，<192.168.0.0>(Area0)和<172.16.0.0>(Area20)路由项已全部过滤。

• 验证区域 20 路由项，以 R8 为例。

```
[R8]display ip routing-table
Route Flags: R - relay, D - download to fib
------------------------------------------------------------------------------
Routing Tables: Public
         Destinations : 19        Routes : 20
  Destination/Mask    Proto   Pre  Cost      Flags NextHop         Interface
        0.0.0.0/0   O_ASE   150  1           D   172.16.1.1      GigabitEthernet0/0/0
      127.0.0.0/8   Direct  0    0           D   127.0.0.1       InLoopBack0
     127.0.0.1/32   Direct  0    0           D   127.0.0.1       InLoopBack0
127.255.255.255/32  Direct  0    0           D   127.0.0.1       InLoopBack0
    172.16.1.0/24   Direct  0    0           D   172.16.1.2      GigabitEthernet0/0/0
    172.16.1.2/32   Direct  0    0           D   127.0.0.1       GigabitEthernet0/0/0
  172.16.1.255/32   Direct  0    0           D   127.0.0.1       GigabitEthernet0/0/0
    172.16.2.0/24   Direct  0    0           D   172.16.2.2      GigabitEthernet0/0/1
    172.16.2.2/32   Direct  0    0           D   127.0.0.1       GigabitEthernet0/0/1
  172.16.2.255/32   Direct  0    0           D   127.0.0.1       GigabitEthernet0/0/1
    172.16.3.0/24   Direct  0    0           D   172.16.3.2      GigabitEthernet2/0/0
    172.16.3.2/32   Direct  0    0           D   127.0.0.1       GigabitEthernet2/0/0
  172.16.3.255/32   Direct  0    0           D   127.0.0.1       GigabitEthernet2/0/0
    172.16.4.0/24   Direct  0    0           D   172.16.4.2      GigabitEthernet1/0/0
    172.16.4.2/32   Direct  0    0           D   127.0.0.1       GigabitEthernet1/0/0
  172.16.4.255/32   Direct  0    0           D   127.0.0.1       GigabitEthernet1/0/0
    172.16.5.0/24   OSPF    10   2           D   172.16.3.1      GigabitEthernet2/0/0
                    OSPF    10   2           D   172.16.4.1      GigabitEthernet1/0/0
    172.16.6.0/24   OSPF    10   2           D   172.16.4.1      GigabitEthernet1/0/0
255.255.255.255/32  Direct  0    0           D   127.0.0.1       InLoopBack0
```

在路由器 R8 上仅存有本区域路由，<192.168.0.0>(Area0)和<10.0.0.0>(Area10)路由项已全部过滤。

• 验证骨干区域 0 路由项，以 R3 和 R2 为例。

```
[R3]display ip routing-table
Route Flags: R - relay, D - download to fib
------------------------------------------------------------------------------
Routing Tables: Public
         Destinations : 22        Routes : 23
  Destination/Mask    Proto   Pre  Cost      Flags NextHop         Interface
        0.0.0.0/0   O_ASE   150  1           D   192.168.4.1     GigabitEthernet0/0/1
                    O_ASE   150  1           D   192.168.3.1     GigabitEthernet0/0/0
       10.0.1.0/24  OSPF    10   2           D   10.0.2.2        GigabitEthernet2/0/0
       10.0.2.0/24  Direct  0    0           D   10.0.2.1        GigabitEthernet2/0/0
       10.0.2.1/32  Direct  0    0           D   127.0.0.1       GigabitEthernet2/0/0
     10.0.2.255/32  Direct  0    0           D   127.0.0.1       GigabitEthernet2/0/0
       10.0.3.0/24  OSPF    10   2           D   10.0.2.2        GigabitEthernet2/0/0
       10.0.4.0/24  OSPF    10   2           D   10.0.2.2        GigabitEthernet2/0/0
       10.0.5.0/24  OSPF    10   3           D   10.0.2.2        GigabitEthernet2/0/0
       10.0.6.0/24  OSPF    10   3           D   10.0.2.2        GigabitEthernet2/0/0
```

```
        127.0.0.0/8  Direct 0    0          D  127.0.0.1       InLoopBack0
        127.0.0.1/32 Direct 0    0          D  127.0.0.1       InLoopBack0
127.255.255.255/32   Direct 0    0          D  127.0.0.1       InLoopBack0
      192.168.3.0/24 Direct 0    0          D  192.168.3.2     GigabitEthernet0/0/0
      192.168.3.2/32 Direct 0    0          D  127.0.0.1       GigabitEthernet0/0/0
    192.168.3.255/32 Direct 0    0          D  127.0.0.1       GigabitEthernet0/0/0
      192.168.4.0/24 Direct 0    0          D  192.168.4.2     GigabitEthernet0/0/1
      192.168.4.2/32 Direct 0    0          D  127.0.0.1       GigabitEthernet0/0/1
    192.168.4.255/32 Direct 0    0          D  127.0.0.1       GigabitEthernet0/0/1
      192.168.5.0/24 Direct 0    0          D  192.168.5.1     GigabitEthernet1/0/0
      192.168.5.1/32 Direct 0    0          D  127.0.0.1       GigabitEthernet1/0/0
    192.168.5.255/32 Direct 0    0          D  127.0.0.1       GigabitEthernet1/0/0
  255.255.255.255/32 Direct 0    0          D  127.0.0.1       InLoopBack0
```

可以看到，路由器 R3 缺少去往<172.16.0.0>和<192.168.1.0>、<192.168.2.0> OSPF 路由条目，因为<172.16.0.0>和<192.168.0.0>OSPF 路由项已被 R1 和 R4 过滤（虽然 R3 与 R1、R4 邻居关系正常）。

```
[R2]display ip routing-table
Route Flags: R - relay, D - download to fib
------------------------------------------------------------------------------
Routing Tables: Public
         Destinations : 23       Routes : 23
  Destination/Mask    Proto  Pre  Cost      Flags NextHop         Interface
          0.0.0.0/0   Static 60   0           RD  202.116.64.2    Serial1/0/0
        127.0.0.0/8   Direct 0    0           D   127.0.0.1       InLoopBack0
        127.0.0.1/32  Direct 0    0           D   127.0.0.1       InLoopBack0
  127.255.255.255/32  Direct 0    0           D   127.0.0.1       InLoopBack0
       172.16.1.0/24  Direct 0    0           D   172.16.1.1      GigabitEthernet2/0/0
       172.16.1.1/32  Direct 0    0           D   127.0.0.1       GigabitEthernet2/0/0
     172.16.1.255/32  Direct 0    0           D   127.0.0.1       GigabitEthernet2/0/0
       172.16.2.0/24  OSPF   10   2           D   172.16.1.2      GigabitEthernet2/0/0
       172.16.3.0/24  OSPF   10   2           D   172.16.1.2      GigabitEthernet2/0/0
       172.16.4.0/24  OSPF   10   2           D   172.16.1.2      GigabitEthernet2/0/0
       172.16.5.0/24  OSPF   10   3           D   172.16.1.2      GigabitEthernet2/0/0
       172.16.6.0/24  OSPF   10   3           D   172.16.1.2      GigabitEthernet2/0/0
      192.168.1.0/24  Direct 0    0           D   192.168.1.1     GigabitEthernet0/0/0
      192.168.1.1/32  Direct 0    0           D   127.0.0.1       GigabitEthernet0/0/0
    192.168.1.255/32  Direct 0    0           D   127.0.0.1       GigabitEthernet0/0/0
      192.168.2.0/24  Direct 0    0           D   192.168.2.1     GigabitEthernet0/0/1
      192.168.2.1/32  Direct 0    0           D   127.0.0.1       GigabitEthernet0/0/1
    192.168.2.255/32  Direct 0    0           D   127.0.0.1       GigabitEthernet0/0/1
    202.116.64.0/24   Direct 0    0           D   202.116.64.1    Serial1/0/0
    202.116.64.1/32   Direct 0    0           D   127.0.0.1       Serial1/0/0
    202.116.64.2/32   Direct 0    0           D   202.116.64.2    Serial1/0/0
  202.116.64.255/32   Direct 0    0           D   127.0.0.1       Serial1/0/0
  255.255.255.255/32  Direct 0    0           D   127.0.0.1       InLoopBack0
```

R2 也缺少去往<10.0.0.0>和<192.168.3.0>、<192.168.3.0>OSPF 路由条目，因为<10.0.0.0>和<192.168.0.0>OSPF 路由项已被过滤（R3 与 R1、R2 邻居关系正常）。

• 连通性验证。

路由器 R10 可以连通公网,因为路由器 R2 有去往<172.16.0.0>的内网路由。

```
[R10]_ping 116.64.64.100
  PING 116.64.64.100: 56  data bytes, press CTRL_C to break
    Reply from 116.64.64.100: bytes=56 Sequence=1 ttl=252 time=30 ms
    Reply from 116.64.64.100: bytes=56 Sequence=2 ttl=252 time=30 ms
    Reply from 116.64.64.100: bytes=56 Sequence=3 ttl=252 time=40 ms
  --- 116.64.64.100 ping statistics ---
    3 packet(s) transmitted
    3 packet(s) received
    0.00% packet loss
    round-trip min/avg/max = 30/33/40 ms
```

路由器 R7 最多只能连通 192.168.1.2,因为路由器 R2 缺少去往<10.0.0.0>的内网路由。

```
[R7]tracert 116.64.64.100
 traceroute to  116.64.64.100(116.64.64.100), max hops: 30 ,packet length: 40,pr
ess CTRL_C to break
 1 10.0.4.2 30 ms  10 ms  10 ms
 2 10.0.1.1 20 ms  30 ms  30 ms
 3 ***
 4 ***
[R7]ping 192.168.1.2
  PING 192.168.1.2: 56 data bytes, press CTRL_C to break
    Reply from 192.168.1.2: bytes=56 Sequence=1 ttl=254 time=30 ms
    Reply from 192.168.1.2: bytes=56 Sequence=2 ttl=254 time=30 ms
    Reply from 192.168.1.2: bytes=56 Sequence=3 ttl=254 time=30 ms
  --- 192.168.1.2 ping statistics ---
    3 packet(s) transmitted
    3 packet(s) received
    0.00% packet loss
    round-trip min/avg/max = 30/30/30 ms
```

工作任务七

配置 OSPF 验证

【工作目的】

理解 md5 和 hmac-md5 认证原理，掌握 OSPF 验证方式和配置过程。

【工作任务】

在工作任务六中，A 公司收购了 B 公司和 C 公司，并在内部划分出 3 个不同区域（公司 A 为 Area0，公司 B 为 Area10，公司 C 为 Area20），通过 OSPF 路由协议互联。企业合并后，某员工想获取原 C 企业主机 3 通信的机密信息，在不影响连通性情况下，将黑客路由器通过交换机 SW1 接入 R4，伪造＜172.16.2.0＞和＜172.16.6.0＞网段，通过 OSPF 路由项欺骗劫持主机 3 流量，从而利用主机 4 冒充并截获发往主机 3 的机密信息。主机 3 发现无法上网，告知管理员后，管理员发现公司 A 和公司 B 去往＜172.16.6.0＞网段经 IP 地址 172.16.2.3 转发，非路由器 R10 的 IP 地址，判断网络中已接入未授权路由器并发起路由项欺骗攻击。为避免再发生类似情况，公司要求对所有路由器开启 OSPF 路由项源端鉴别功能，对区域间路由器采用 OSPF 链路认证，对区域内路由器采用 OSPF 区域认证，从而丢弃未通过源端鉴别的 OSPF 路由分组信息，具体任务如下。

（1）模拟 hacker 路由器接入 R4，伪造＜172.16.2.0＞和＜172.16.6.0＞网段分组信息。

（2）验证主机 4 截获主机 2 与主机 3、主机 1 和主机 3 流量。

（3）对区域间路由器采用 OSPF 链路认证，采用更为安全的 hmac-md5 认证算法，认证密钥为 huawei。

（4）对区域内路由器采用 OSPF 区域认证。由于区域内路由器性能一般，采用 md5 认证算法，其中 Area0 认证密钥为 gdcp，Area10 认证密钥为 gdcp10，Area20 认证密钥为 gdcp20。

（5）配置源端鉴别后，验证 hacker 路由器无法接入网络。

【工作背景】

OSPF（open shortest path first，开放式最短路径优先）是链路状态协议，通过组播 LSA 链路状态通告信息建立链路状态数据库，生成最短路径树。假如攻击者伪造链路状态 LSA 分组报文和最小距离开销，便可诱导其他路由器更新最短通路，将流量引至攻击者，从而劫持用户会话。

为防范 OSPF 路由项欺骗攻击，可在路由器上启用 OSPF 源端鉴别功能，从接口收到的 LSA 分组只有成功通过源端鉴别后，才能提交给本地 OSPF 进程处理，避免未授权路由器发起的 OSPF 路由项欺骗攻击。

【任务分析】

OSPF 源端鉴别支持区域认证和链路认证两种认证方式。

(1) 区域认证。区域内所有路由器认证模式、密钥和密钥标识符必须一致。

(2) 链路认证。一条链路上双方路由器接口的认证模式、密钥和密钥标识符必须一致。

注意：OSPF 链路认证比区域认证更加灵活，可对不同链路单独设置不同密码。如果路由器同时配置了区域认证和链路认证，优先使用链路认证建立 OSPF 邻居。

为防范 OSPF 路由项欺骗攻击，可通过 md5、hmac-md5 认证算法对邻居节点发送的 OSPF 路由分组报文进行认证(包含身份认证和报文完整性认证)，如图 7-1 所示。以 hmac-md5 算法为例，路由器 R1 和 R2 相邻路由器由管理员配置相同的共享密钥 huawei。R1 在通告 OSPF 报文前，基于 huawei 密钥和 hmac-md5 摘要算法(hmac-md5 是一种改进 md5 密钥完整性验证方法，算法不可逆向推导，比 md5 算法复杂，安全性更高，但会牺牲路由器部分性能)将自身 OSPF 路由分组生成 256 位鉴别码，连同源 OSPF 路由分组经 R1 的 GE 0/0/0 接口发送至 R2。R2 收到来自邻居 R1 发过来的 OSPF 路由分组和摘要信息，为判断 OSPF 路由分组的可靠性，同样通过 huawei 密钥和 hmac-md5 摘要算法生成鉴别码，并与从 R1 收到的鉴别码进行匹配，如一致则接收，不一致则丢弃，从而保证：①收到的 OSPF 路由分组信息在途中不被篡改(完整性验证)；②鉴别路由器 R1 身份(对方密钥一定是 huawei，因为生成的鉴别码一致，摘要算法一致，则密钥肯定一致)。

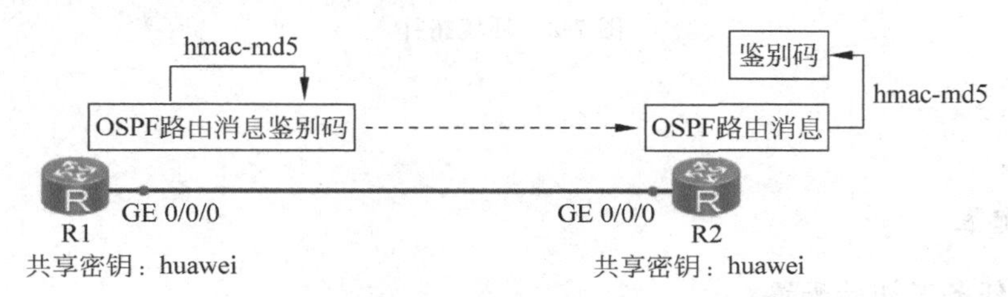

图 7-1 OSPF 认证过程

【设备器材】

交换机(S5700 或者 S3700)1 台，无须配置；路由器(AR1220)12 台，需要添加 1GEC 千兆接口模块或 2SA 串口模块。

主机 5 台，承担角色见表 7-1。

表 7-1 主机配置表

角　　色	接入方式	IP 地址	所属公司
主机 1	eNSP PC 接入	192.168.5.10/24	公司 A
主机 2	eNSP PC 接入	10.0.6.10/24	公司 B
主机 3	eNSP PC 接入	172.16.6.10/24	公司 C
主机 4	eNSP PC 接入	172.16.6.10/24	黑客计算机
公网服务器	eNSP Server 接入	116.64.64.100	

【环境拓扑】

环境拓扑如图 7-2 所示。

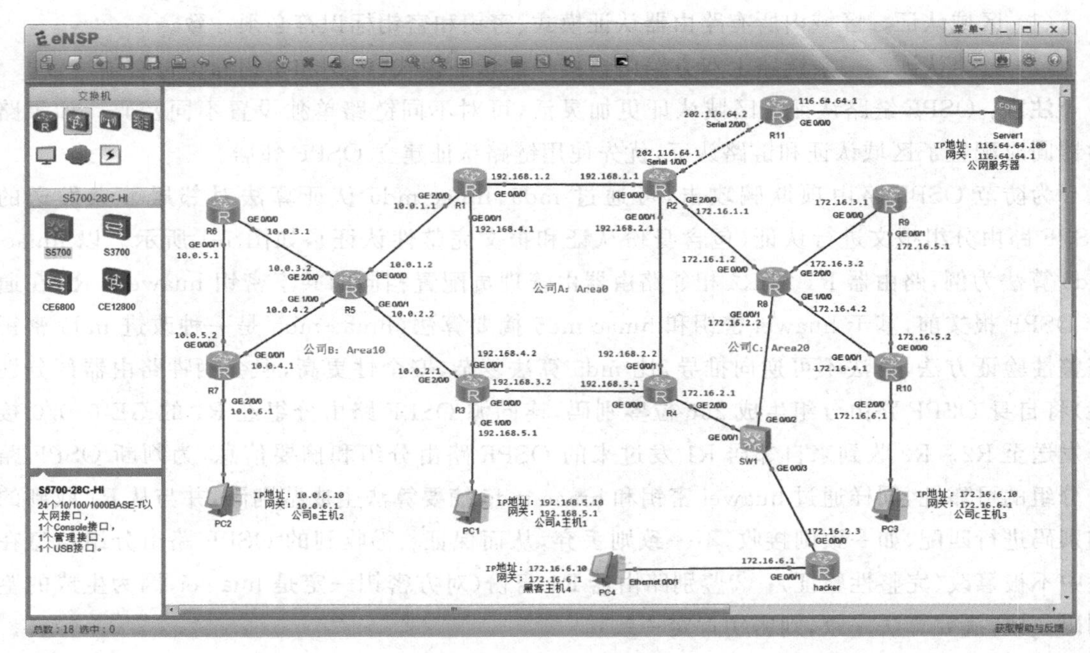

图 7-2　环境拓扑

【工作过程】

1. 基本配置

(1) 工作任务五初始配置。

请读者根据工作任务五网络拓扑完成 OSPF 区域间互联，实现主机 1、主机 2 和主机 3 内外网互通，如图 7-3 所示。

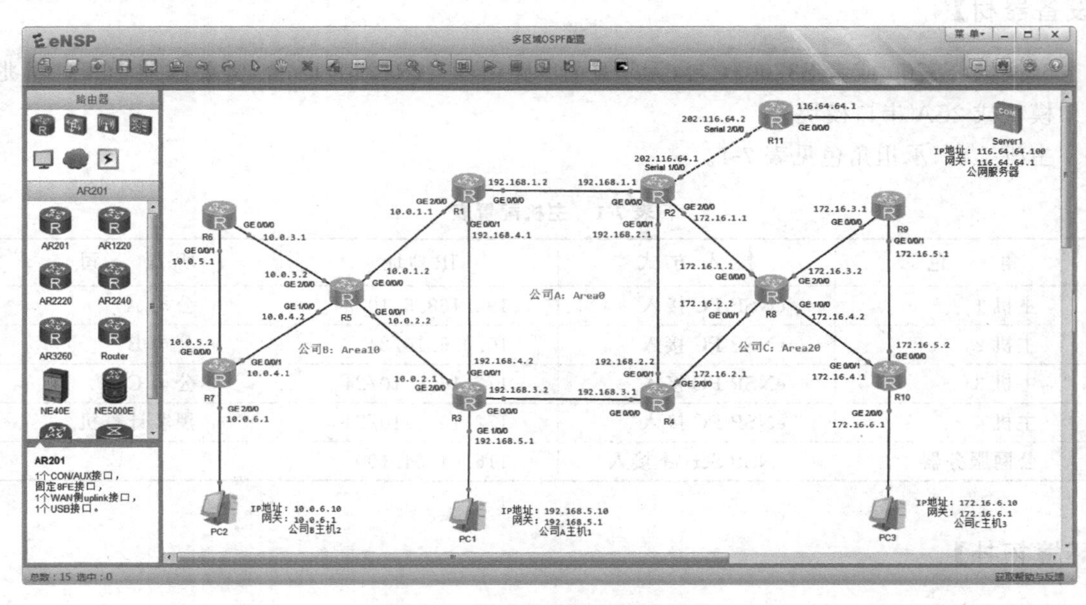

图 7-3　原工作任务五网络拓扑

(2) 将 hacker 路由器接入 R4，伪造<172.16.2.0>和<172.16.6.0>网段分组信息。

根据环境拓扑(见图 7-2)将 hacker 路由器接入 R4，并做如下配置。

```
[Huawei]sysname hacker
[hacker]interface GigabitEthernet 0/0/0
[hacker-GigabitEthernet0/0/0]ip address 172.16.2.3 24
[hacker-GigabitEthernet0/0/0]quit
[hacker]interface GigabitEthernet 0/0/1
[hacker-GigabitEthernet0/0/1]ip address 172.16.6.1 24
[hacker-GigabitEthernet0/0/1]quit
[hacker]ospf 1
[hacker-ospf-1]area 20
[hacker-ospf-1-area-0.0.0.20]network 172.16.2.0 0.0.0.255
[hacker-ospf-1-area-0.0.0.20]network 172.16.6.0 0.0.0.255
[hacker-ospf-1-area-0.0.0.20]quit
[hacker-ospf-1]quit
[hacker]
```

(3) 验证通过主机 4 截获主机 2 与主机 3、主机 1 和主机 3 流量。

① 查看路由器 R7 路由表。由于篇幅限制，以下路由表不列举直连路由条目。

```
[R7]display ip routing-table
Route Flags: R - relay, D - download to fib
------------------------------------------------------------------------------
Routing Tables: Public
         Destinations : 28        Routes : 29
  Destination/Mask   Proto  Pre  Cost      Flags NextHop         Interface
        0.0.0.0/0   O_ASE  150  1           D   10.0.4.2        GigabitEthernet0/0/1
      10.0.1.0/24   OSPF   10   2           D   10.0.4.2        GigabitEthernet0/0/1
      10.0.2.0/24   OSPF   10   2           D   10.0.4.2        GigabitEthernet0/0/1
      10.0.3.0/24   OSPF   10   2           D   10.0.5.1        GigabitEthernet0/0/0
                    OSPF   10   2           D   10.0.4.2        GigabitEthernet0/0/1
    172.16.1.0/24   OSPF   10   4           D   10.0.4.2        GigabitEthernet0/0/1
    172.16.2.0/24   OSPF   10   4           D   10.0.4.2        GigabitEthernet0/0/1
    172.16.3.0/24   OSPF   10   5           D   10.0.4.2        GigabitEthernet0/0/1
    172.16.4.0/24   OSPF   10   5           D   10.0.4.2        GigabitEthernet0/0/1
    172.16.5.0/24   OSPF   10   6           D   10.0.4.2        GigabitEthernet0/0/1
    172.16.6.0/24   OSPF   10   5           D   10.0.4.2        GigabitEthernet0/0/1
   192.168.1.0/24   OSPF   10   3           D   10.0.4.2        GigabitEthernet0/0/1
   192.168.2.0/24   OSPF   10   4           D   10.0.4.2        GigabitEthernet0/0/1
   192.168.3.0/24   OSPF   10   3           D   10.0.4.2        GigabitEthernet0/0/1
   192.168.4.0/24   OSPF   10   3           D   10.0.4.2        GigabitEthernet0/0/1
   192.168.5.0/24   OSPF   10   3           D   10.0.4.2        GigabitEthernet0/0/1
```

在原工作任务五中，通过路由器 R7 查看到去往<172.16.6.0>网段路由条目如下：

```
172.16.6.0/24       OSPF   10   6           D   10.0.4.2        GigabitEthernet0/0/1
```

由于 hacker 路由器伪造<172.16.6.0>网段分组，对于路由器 R7 来说去往<172.16.6.0>网段拥有更小的路径开销(开销值从 6 变为 5)，从而更新路由表，通过 hacker 路由器去往<172.16.6.0>网段。

② 主机 1、主机 2 与 IP 地址为 172.16.6.10 的主机的通路测试。

在主机 1、主机 2 上分别运行命令 tracert 172.16.6.10，结果如图 7-4 所示，可以看到去往 IP 地址为 172.16.6.10 的主机都通过 IP 地址为 172.16.2.3 hacker 路由器转发。

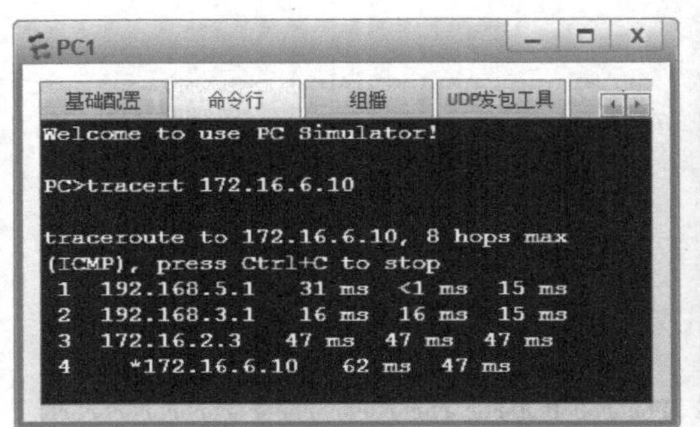

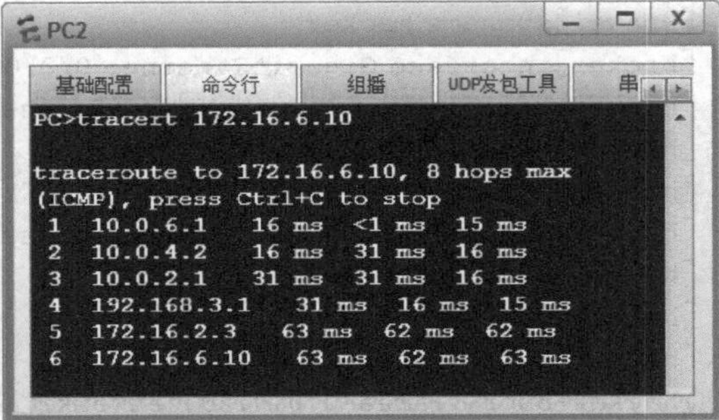

图 7-4 通路测试图

③ 抓包测试。打算利用主机 1 与主机 2 与主机 3 之间进行通信，在主机命令行窗口输入 ping 172.16.6.10 -t。在 hacker 路由器的 G0/0/0 或 G0/0/1 接口启用抓包程序，进入 Wireshark 界面，截获到来自主机 1(192.168.5.10)和主机 2(10.0.6.10)发往主机 4(172.16.6.10)的 ICMP 包，如图 7-5 所示。

No.	Time	Source	Destination	Protocol
141	31.387000	172.16.2.1	224.0.0.5	OSPF
142	31.652000	HuaweiTe_17:33:e9	Spanning-tree-(for-…	STP
143	31.808000	192.168.5.10	172.16.6.10	ICMP
144	31.824000	172.16.6.10	192.168.5.10	ICMP
145	32.089000	10.0.6.10	172.16.6.10	ICMP
146	32.089000	172.16.6.10	10.0.6.10	ICMP
147	32.854000	192.168.5.10	172.16.6.10	ICMP
148	32.869000	172.16.6.10	192.168.5.10	ICMP
149	32.978000	172.16.2.2	224.0.0.5	OSPF
150	33.150000	10.0.6.10	172.16.6.10	ICMP
151	33.150000	172.16.6.10	10.0.6.10	ICMP

图 7-5 hacker 路由器截获来自主机 1 和主机 2 的数据包

(4) 区域间公司路由器源端鉴别配置。

采用 hmac-md5 认证算法，密钥为 huawei。

① 对区域间路由器启用 OSPF 路由项源端鉴别功能，鉴别方式采用链路认证，涉及 4 条链路，分别是 R1 和 R5、R3 和 R5、R2 和 R8、R4 和 R8。

```
[R1]interface GigabitEthernet 2/0/0
[R1-GigabitEthernet2/0/0]ospf authentication-mode hmac-md5 15 cipher huawei
```

//• hmac-md5 参数：hmac-md5 是 hmac 算法的一个特例，用 md5 作为 hmac 的 Hash 函数，算法不可逆向推导，通过信息摘要方式保证数据完整性(不被篡改)。10 为密钥标识符，范围<1~255>。只有密钥相同，如本例中的 huawei，密钥标识符也相同，才能建立邻居关系。利用密钥标识符可解决多个密码不好记，相同密码不安全的问题。为方便记忆，密钥标识符采用 15，标识 R1 和 R5 之间认证关系。OSPF 邻居认证的安全性主要由密钥长度决定，与密钥标识符关系不大；

- md5 参数：md5 摘要算法同样是一种基于密钥的报文完整性验证方法，算法不可逆向推导；
- cipher 参数：导出的设备配置信息或 display 信息中将有关 huawei 的密钥字符进行加密处理。注意，cipher 参数可以不写，默认字符须加密；
- plain 参数：不加密，配置信息中 huawei 密钥直接可见，不安全

```
[R1-GigabitEthernet2/0/0]quit
[R1]

[R5]interface GigabitEthernet 0/0/0
[R5-GigabitEthernet0/0/0]ospf authentication-mode hmac-md5 15 plain huawei
```

//R1 和 R5 密钥和密钥标识符必须一致，否则无法通过源端鉴别。但是究竟是 cipher 加密保存在本地，还是 plain 保存在本地，可由双方路由器自行决定。为让读者理解 cipher 参数与 plain 参数区别，R5 采用 plain 参数，但一般情况下应采用 cipher 参数保证安全性

```
[R5-GigabitEthernet0/0/0]quit
[R5]
```

② 通过命令 display current-configuration 查看当前配置。

```
[R1]display current-configuration
[V200R003C00]
 sysname R1
 board add 0/2 1GEC
 snmp-agent local-engineid 800007DB03000000000000
 snmp-agent
 ...
#
interface GigabitEthernet2/0/0
 ip address 10.0.1.1 255.255.255.0
 ospf authentication-mode hmac-md5 15 cipher %$%$Y4ZL#NxI7)[O})-:rT10b%#o%$%$
#
```

路由器 R1 采用 cipher 参数，密钥 huawei 经加密处理（密钥标识符 15 不加密）后，无法直接查看。如忘记密码，只能在双方路由器接口下重新设置新密码加以解决。

```
[R5]display current-configuration
#
 sysname R5
#
 board add 0/1 1GEC
 board add 0/2 1GEC
#
 snmp-agent local-engineid 800007DB03000000000000
 snmp-agent
...
#
interface GigabitEthernet0/0/0
 ip address 10.0.1.2 255.255.255.0
 ospf authentication-mode hmac-md5 15 plain huawei
#
```

路由器 R5 采用 plain 参数，密钥 huawei 可直接呈现，会产生安全问题。

③ 继续配置区域间路由器源端鉴别功能。

```
[R3]interface GigabitEthernet 2/0/0
[R3-GigabitEthernet2/0/0]ospf authentication-mode hmac-md5 35 cipher huawei
[R3-GigabitEthernet2/0/0]quit
[R3]
-------------------------------------------------------------------------------
[R5]interface GigabitEthernet 0/0/1
[R5-GigabitEthernet0/0/1]ospf authentication-mode hmac-md5 35 cipher huawei
[R5-GigabitEthernet0/0/1]quit
[R5]
-------------------------------------------------------------------------------
[R2]interface GigabitEthernet 2/0/0
[R2-GigabitEthernet2/0/0]ospf authentication-mode hmac-md5 28 cipher huawei
[R2-GigabitEthernet2/0/0]quit
[R2]
-------------------------------------------------------------------------------
[R8]interface GigabitEthernet 0/0/0
[R8-GigabitEthernet0/0/0]ospf authentication-mode hmac-md5 28 cipher huawei
[R8-GigabitEthernet0/0/0]quit
[R8]
-------------------------------------------------------------------------------
[R4]interface GigabitEthernet 2/0/0
[R4-GigabitEthernet2/0/0]ospf authentication-mode hmac-md5 48 cipher huawei
[R4-GigabitEthernet2/0/0]quit
[R4]
-------------------------------------------------------------------------------
[R8]interface GigabitEthernet 0/0/1
[R8-GigabitEthernet0/0/1]ospf authentication-mode hmac-md5 48 cipher huawei
[R8-GigabitEthernet0/0/1]quit
[R8]
```

上述配置完成后，即使将 hacker 路由器物理上接入路由器 R4，也会因为伪造的路由项＜172.16.2.0＞和＜172.16.6.0＞无法通过路由器 R4 和路由器 R8 源端鉴别而被丢弃，从而保证 OSPF 邻居间路由项信息的安全性。

（5）区域内公司路由器源端鉴别配置。

虽然上述配置已达到防范 hacker 路由器接入的目的，但是管理员为避免区域内发生类似事件，建议在路由器 OSPF 区域配置认证（采用在接口配置 OSPF 认证的方式操作上较为烦琐），采用 md5 认证算法。为方便密钥标识和管理，其中：

① Area0 认证密钥为 gdcp，密钥标识符 1（密钥标识符不能为 0）。

② Area10 认证密钥为 gdcp10，密钥标识符 10。

③ Area20 认证密钥为 gdcp20，密钥标识符 20。

```
[R1]ospf 1
[R1-ospf-1]area 0
[R1-ospf-1-area-0.0.0.0]authentication-mode md5 1 cipher gdcp
[R1-ospf-1-area-0.0.0.0]quit
[R1-ospf-1]area 10
[R1-ospf-1-area-0.0.0.10]authentication-mode md5 10 cipher gdcp10
[R1-ospf-1-area-0.0.0.10]quit
```

```
[R1-ospf-1]quit
[R1]
```

```
[R2-ospf-1]
[R2-ospf-1]area 0
[R2-ospf-1-area-0.0.0.0]authentication-mode md5 1 cipher gdcp
[R2-ospf-1-area-0.0.0.0]quit
[R2-ospf-1]area 20
[R2-ospf-1-area-0.0.0.20]authentication-mode md5 20 cipher gdcp20
[R2-ospf-1-area-0.0.0.20]quit
[R2-ospf-1]quit
[R2]
```

```
[R3]ospf 1
[R3-ospf-1]area 0
[R3-ospf-1-area-0.0.0.0]authentication-mode md5 1 cipher gdcp
[R3-ospf-1-area-0.0.0.0]quit
[R3-ospf-1]area 10
[R3-ospf-1-area-0.0.0.10]authentication-mode md5 10 cipher gdcp10
[R3-ospf-1-area-0.0.0.10]quit
[R3-ospf-1]quit
[R3]
```

```
[R4]ospf 1
[R4-ospf-1]area 0
[R4-ospf-1-area-0.0.0.0]authentication-mode md5 1 cipher gdcp
[R4-ospf-1-area-0.0.0.0]quit
[R4-ospf-1]area 20
[R4-ospf-1-area-0.0.0.20]authentication-mode md5 20 cipher gdcp20
[R4-ospf-1-area-0.0.0.20]quit
[R4-ospf-1]quit
[R4]
```

```
[R5]ospf 1
[R5-ospf-1]area 10
[R5-ospf-1-area-0.0.0.10]authentication-mode md5 10 cipher gdcp10
[R5-ospf-1-area-0.0.0.10]quit
[R5-ospf-1]quit
[R5]
```

路由器 R6 和路由器 R7 配置与路由器 R5 相似，请读者自行配置。

```
[R8]ospf 1
[R8-ospf-1]area 20
[R8-ospf-1-area-0.0.0.20]authentication-mode md5 20 cipher gdcp20
[R8-ospf-1-area-0.0.0.20]quit
[R8-ospf-1]quit
[R8]
```

路由器 R9 和路由器 R10 配置与路由器 R8 相似，请读者自行配置。

2. 任务验证

(1) 主机连通性验证。

主机1、主机2和主机3之间能够相互连通，并能ping通公网服务器。其中主机2测试如图7-6所示。

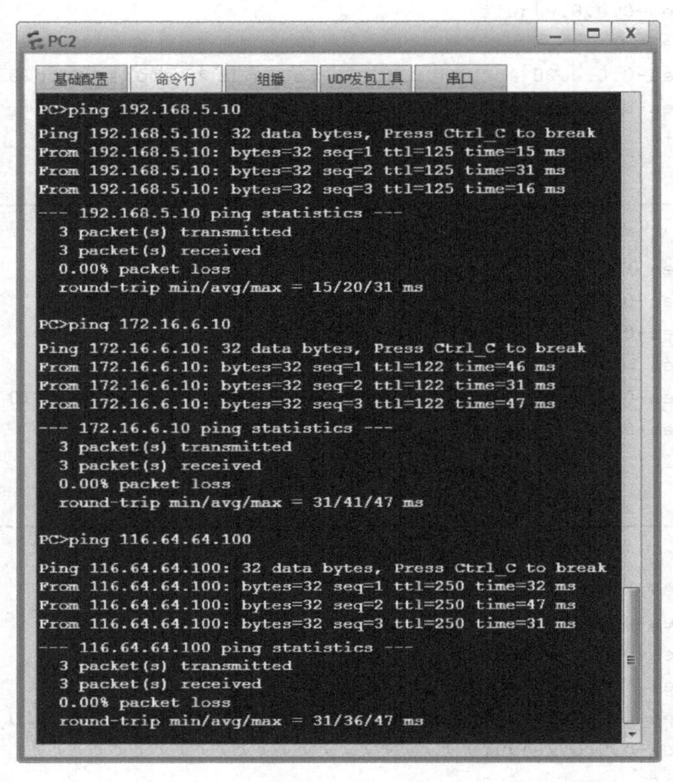

图7-6 主机2可以连通主机1、主机3和公网服务器

(2) 主机1、主机2与IP地址为172.16.6.10的主机通路测试。

在主机1、主机2上分别运行命令tracert 172.16.6.10，结果如图7-7所示，可以看到去往IP地址为172.16.6.10的数据都通过IP地址为172.16.4.1的路由器R10转发，不再经过hacker路由器。

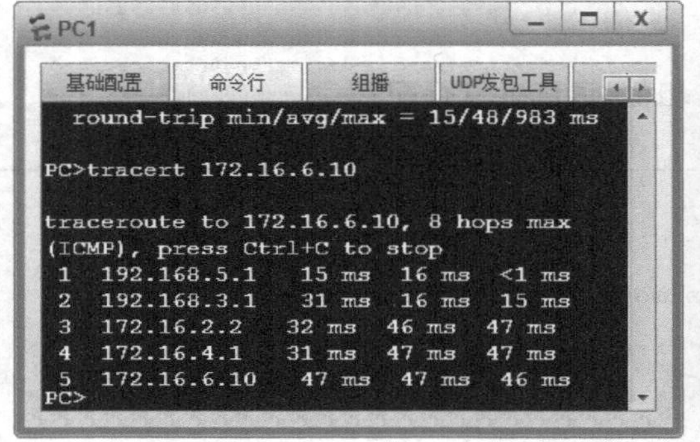

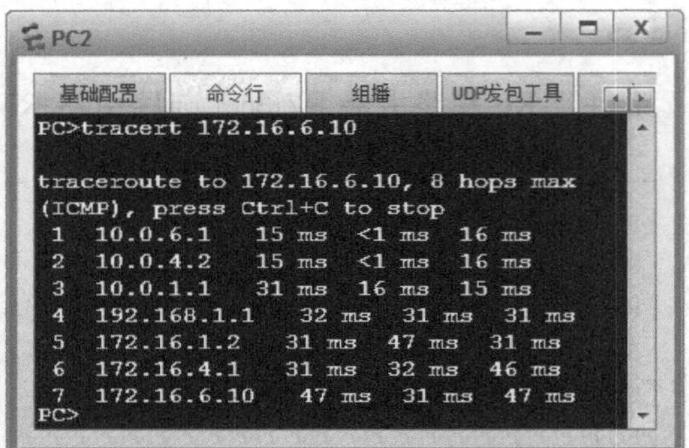

图7-7 通路测试图

(3) 路由条目验证。

以路由器 R7 为例查看路由表。由于篇幅限制,在以下路由表中不列举直连路由条目。

```
[R7]display ip routing-table
Route Flags: R - relay, D - download to fib
------------------------------------------------------------------------------
Routing Tables: Public
         Destinations : 28        Routes : 29
  Destination/Mask    Proto   Pre  Cost      Flags NextHop         Interface
        0.0.0.0/0   O_ASE   150  1           D   10.0.4.2        GigabitEthernet0/0/1
      10.0.1.0/24   OSPF    10   2           D   10.0.4.2        GigabitEthernet0/0/1
      10.0.2.0/24   OSPF    10   2           D   10.0.4.2        GigabitEthernet0/0/1
      10.0.3.0/24   OSPF    10   2           D   10.0.5.1        GigabitEthernet0/0/0
                    OSPF    10   2           D   10.0.4.2        GigabitEthernet0/0/1
    172.16.1.0/24   OSPF    10   4           D   10.0.4.2        GigabitEthernet0/0/1
    172.16.2.0/24   OSPF    10   5           D   10.0.4.2        GigabitEthernet0/0/1
    172.16.3.0/24   OSPF    10   5           D   10.0.4.2        GigabitEthernet0/0/1
    172.16.4.0/24   OSPF    10   5           D   10.0.4.2        GigabitEthernet0/0/1
    172.16.5.0/24   OSPF    10   6           D   10.0.4.2        GigabitEthernet0/0/1
    172.16.6.0/24   OSPF    10   6           D   10.0.4.2        GigabitEthernet0/0/1
   192.168.1.0/24   OSPF    10   3           D   10.0.4.2        GigabitEthernet0/0/1
   192.168.2.0/24   OSPF    10   4           D   10.0.4.2        GigabitEthernet0/0/1
   192.168.3.0/24   OSPF    10   5           D   10.0.4.2        GigabitEthernet0/0/1
   192.168.4.0/24   OSPF    10   3           D   10.0.4.2        GigabitEthernet0/0/1
   192.168.5.0/24   OSPF    10   6           D   10.0.4.2        GigabitEthernet0/0/1
```

可以看到路由器 R7 有去往全网的 OSPF 路由条目,其中去往<172.16.6.0>网段不再经过 hacker 路由器,网段开销值从 5 变为 6。

【任务总结】

(1) 如果在路由器上同时配置了区域认证和链路认证,优先使用链路认证建立 OSPF 邻居。

(2) 在配置 OSPF 路由项源端鉴别时,相邻路由器之间接口必须具备相同的鉴别方式(如 md5 或 hmac-md5)、相同的鉴别密钥(密钥存储方式可以不同,如 cipher 或者 plain)和相同的密钥标识符,否则不能建立邻居关系。

工作任务八

OSPF 虚链路

【工作目的】

理解 OSPF 虚链路应用场景,掌握 OSPF 虚链路配置过程。

【工作任务】

公司 A(Area0)收购公司 B(Area10),通过区域间 OSPF 路由协议互联。某日,由于路由器 R2 和路由器 R3 之间链路故障,导致骨干区域 Area0 被分割成两个逻辑区域,原因未明。由于路由器 R1 为非骨干区域边界路由器,不能转发区域间路由,因此主机 1 无法连通内网服务器和公网。公司采取应急响应,要求在路由器 R2 和路由器 R3 上配置虚链路,连接 Area0 两个逻辑区域,暂时恢复网络连通性,具体任务如下。

(1) 根据拓扑图(见图 8-2)配置区域间 OSPF 路由协议,主机 1 能够访问内网服务器 Web 服务。

(2) 在路由器 R3 上配置 Easy IP,让主机 1 能够访问公网主机 2。

(3) 在路由器 R3 上配置 Nat Server,公网主机 2 可以访问内网服务器 Web 服务。

(4) 断开路由器 R2 和路由器 R3 之间链路,模拟线路故障。

(5) 在路由器 R2 和路由器 R3 上配置 OSPF 虚链路,恢复网络连通性,实现子任务 2 和子任务 3 要求。

【工作背景】

OSPF 区域化设计采用了 Hub-Spoke 的架构,所有区域中定义出一个核心区域(骨干区域),其他区域必须与核心区域(骨干区域)相连,骨干区域边界路由器(连接 Area0 骨干区域和其他非骨干区域的路由器)负责通告区域间 OSPF 路由。

然而在某些情况下,由于各种条件限制,常规区域无法与骨干区域直连,而是连接至其他常规区域,导致无法获得其他区域间路由信息(处于常规区域的边界路由器无法转发区域间信息)。如图 8-1 所示,处于区域 Area20 的路由器无法获得 Area0 区域路由信息。

还有一种情况,OSPF 要求骨干区域必须唯一且是连续的。然而由于链路故障,骨干区域被分割成多个子区域。同样由于处于常规区域的边界路由器无法转发区域间信息,导致这些区域之间相互独立,无法连通。

为解决这些问题,OSPF 允许创建虚链路(virtual link)将多个不能直接与骨干区域相连的区域逻辑上连接在一起,从而扩展常规区域地理范畴。

【任务分析】

在配置 OSPF 虚链路时,需要给两端路由器配置 Router-ID 以进行标识。对于 Router-ID

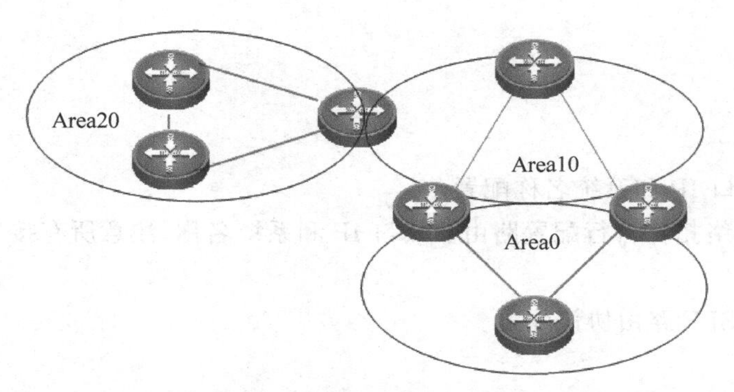

图 8-1　Area20 不能与 Area0 连接无法获得区域间路由通告

不能随意设置，虚链路通过 Router-ID 指定的 IP 进行寻址。如双方 IP 无法通信，则无法建立虚链路。

虚链路在网络中会穿越其他区域。为提高安全性，通常情况下建议对虚链路进行认证。

【设备器材】

路由器(AR1220)6 台。

主机 3 台，承担角色见表 8-1。

表 8-1　主机配置表

角　　色	接 入 方 式	IP 地址	所 属 公 司
主机 1	eNSP Client 接入	172.16.3.10/24	公司 A
主机 2	eNSP Client 接入	116.64.64.100	公网
内网 Web 服务器	eNSP Server 接入	172.16.4.100/24	公司 A

【环境拓扑】

环境拓扑如图 8-2 所示。

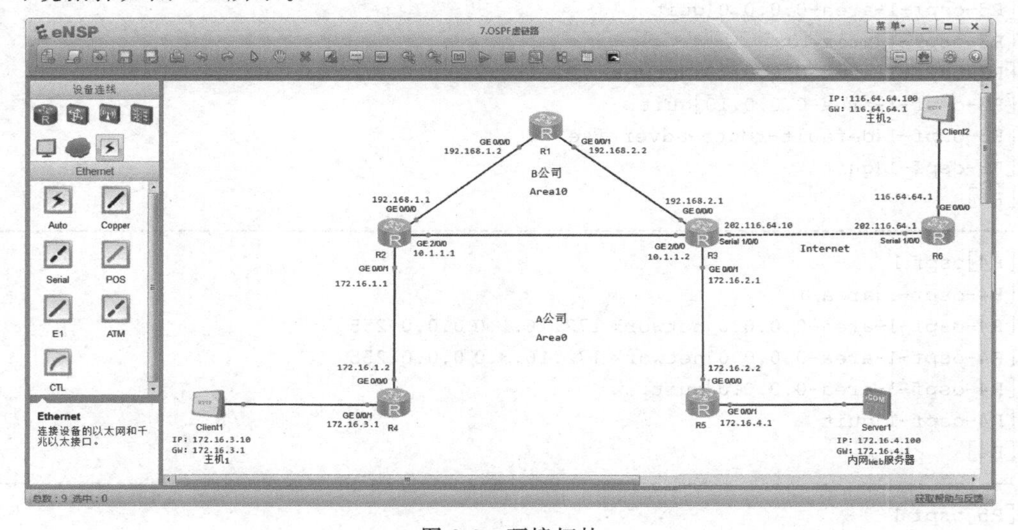

图 8-2　环境拓扑

【工作过程】

1. 基本配置

(1) 路由器接口 IP 和系统名称配置。

请读者根据网络拓扑自行配置路由器接口 IP 和系统名称,注意所有接口子网掩码长度均为 24。

(2) 区域间 OSPF 路由协议配置。

```
[R1]ospf 1
[R1-ospf-1]area 10
[R1-ospf-1-area-0.0.0.10]network 192.168.1.0 0.0.0.255
[R1-ospf-1-area-0.0.0.10]network 192.168.2.0 0.0.0.255
[R1-ospf-1-area-0.0.0.10]quit
[R1-ospf-1]quit
[R1]
-----------------------------------------------------------------------------------------
[R2]ospf 1
[R2-ospf-1]area 0
[R2-ospf-1-area-0.0.0.0]network 10.1.1.0 0.0.0.255
[R2-ospf-1-area-0.0.0.0]network 172.16.1.0 0.0.0.255
[R2-ospf-1-area-0.0.0.0]quit
[R2-ospf-1]area 10
[R2-ospf-1-area-0.0.0.10]network 192.168.1.0 0.0.0.255
[R2-ospf-1-area-0.0.0.10]quit
[R2-ospf-1]quit
[R2]
-----------------------------------------------------------------------------------------
[R3]ospf 1
[R3-ospf-1]area 0
[R3-ospf-1-area-0.0.0.0]network 10.1.1.0 0.0.0.255
[R3-ospf-1-area-0.0.0.0]network 172.16.2.0 0.0.0.255
[R3-ospf-1-area-0.0.0.0]quit
[R3-ospf-1]area 10
[R3-ospf-1-area-0.0.0.10]network 192.168.2.0 0.0.0.255
[R3-ospf-1-area-0.0.0.10]quit
[R3-ospf-1]default-route-advertise
[R3-ospf-1]quit
[R3]
-----------------------------------------------------------------------------------------
[R4]ospf 1
[R4-ospf-1]area 0
[R4-ospf-1-area-0.0.0.0]network 172.16.1.0 0.0.0.255
[R4-ospf-1-area-0.0.0.0]network 172.16.3.0 0.0.0.255
[R4-ospf-1-area-0.0.0.0]quit
[R4-ospf-1]quit
[R4]
-----------------------------------------------------------------------------------------
[R5]ospf 1
[R5-ospf-1]area 0
```

```
[R5-ospf-1-area-0.0.0.0]network 172.16.2.0 0.0.0.255
[R5-ospf-1-area-0.0.0.0]network 172.16.4.0 0.0.0.255
[R5-ospf-1-area-0.0.0.0]quit
[R5-ospf-1]quit
[R5]
```

(3) 在路由器 R3 上配置默认路由、Easy-IP 和 nat server(发布内网 Web 服务器)。

```
[R3]ip route-static 0.0.0.0 0.0.0.0 202.116.64.1
[R3]acl 2000
[R3-acl-basic-2000]rule permit source any
[R3-acl-basic-2000]quit
[R3]interface Serial 1/0/0
[R3-Serial1/0/0]nat outbound 2000
[R3-Serial1/0/0]quit
[R3]interface Serial 1/0/0
[R3-Serial1/0/0]nat server protocol tcp global current-interface 80 inside 172.16.4.
100 80
[R3-Serial1/0/0]quit
[R3]
```

(4) 在 Server1 发布公司 Web 站点。

下载并解压"公司 Web 服务器站点"压缩包。单击 Server1 文件图标打开 Server1 窗口,在 Server1 窗口选中"服务器信息"选项卡,选择 HttpServer 选项指定解压的站点根目录路径,单击"启动"按钮开启 80 端口,如图 8-3 所示。

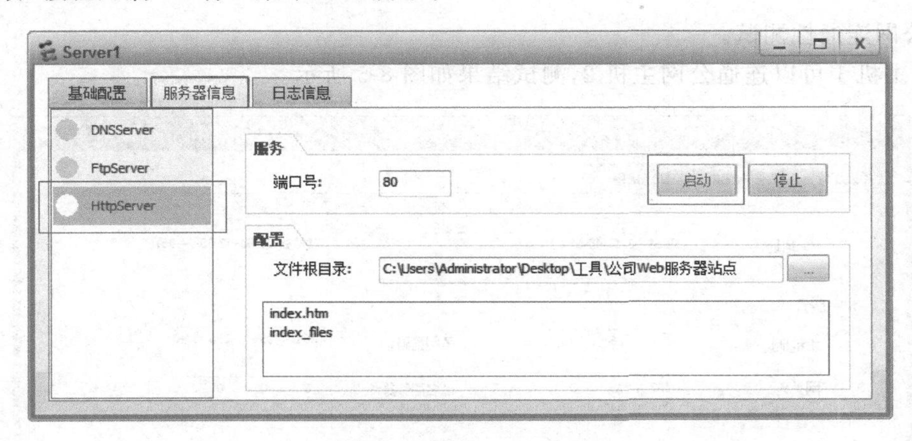

图 8-3 配置 HttpServer

(5) 连通性测试。

① 查看路由器 R4 路由表。

由于篇幅限制,以下路由表中不列举直连路由条目。

```
[R4]display ip routing-table
Route Flags: R - relay, D - download to fib
------------------------------------------------------------------------------
Routing Tables: Public
         Destinations : 16        Routes : 16
  Destination/Mask   Proto  Pre  Cost      Flags NextHop         Interface
        0.0.0.0/0    O_ASE  150  1           D   172.16.1.1      GigabitEthernet0/0/0
```

```
     10.1.1.0/24  OSPF   10  2       D  172.16.1.1     GigabitEthernet0/0/0
   172.16.2.0/24  OSPF   10  3       D  172.16.1.1     GigabitEthernet0/0/0
   172.16.4.0/24  OSPF   10  4       D  172.16.1.1     GigabitEthernet0/0/0
  192.168.1.0/24  OSPF   10  2       D  172.16.1.1     GigabitEthernet0/0/0
  192.168.2.0/24  OSPF   10  3       D  172.16.1.1     GigabitEthernet0/0/0
```

由路由表可以看到，路由器 R4 拥有内网所有网段 OSPF 路由条目。

② 内网 Web 服务器访问测试。

Server1 配置完成后，内网主机 1 通过地址 http://172.16.4.100/index.htm 可以访问 Server1 发布的 Web 站点；公网主机 2 通过地址 http://202.116.64.10/index.htm 可以访问 Server1 发布的 Web 站点，如图 8-4 所示。

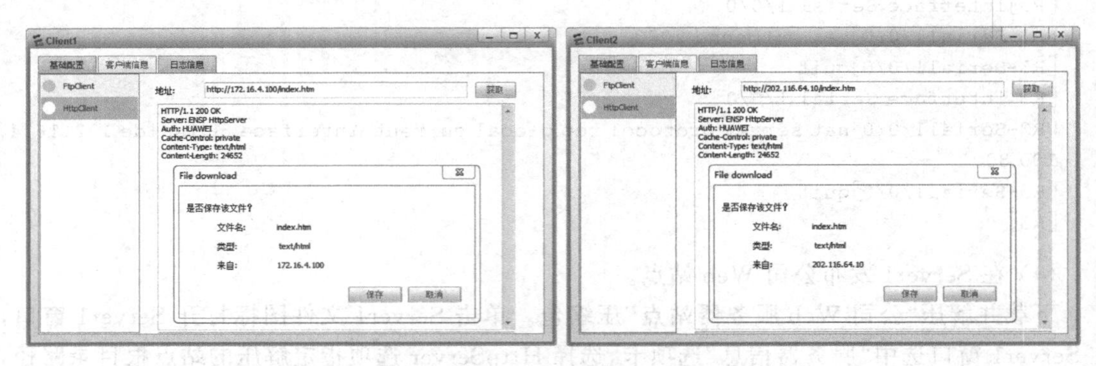

图 8-4 主机 1、主机 2 可以访问 Server1 Web 站点

③ 公网连通性测试。

内网主机 1 可以连通公网主机 2，测试结果如图 8-5 所示。

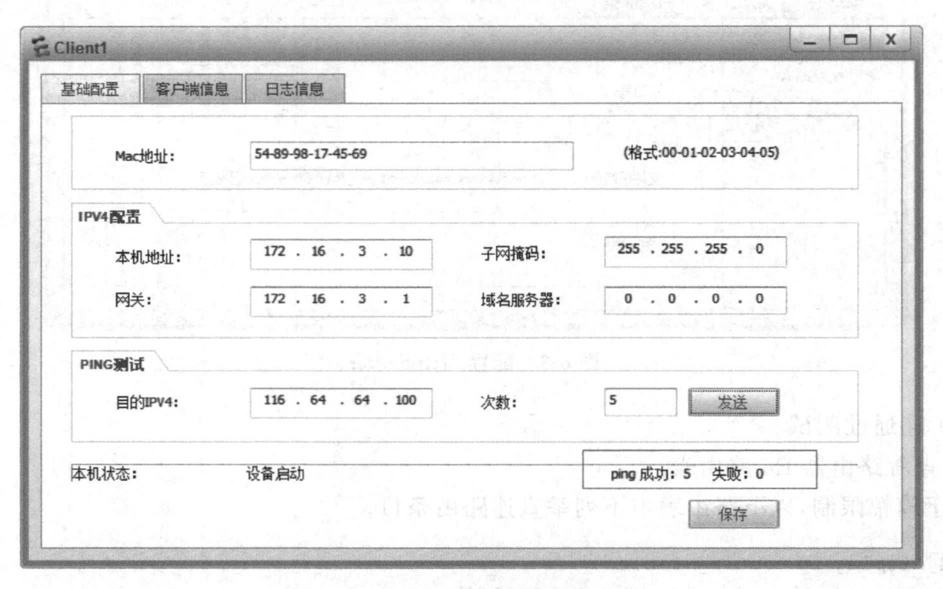

图 8-5 主机 1 可以连通公网主机 2

(6) A 公司主链路断开测试。

断开路由器 R2 与路由器 R3 之间链路，模拟线路故障。此时 Area0 骨干区域被分割成两个逻辑区域，分别与 Area10 连接，如图 8-6 所示。

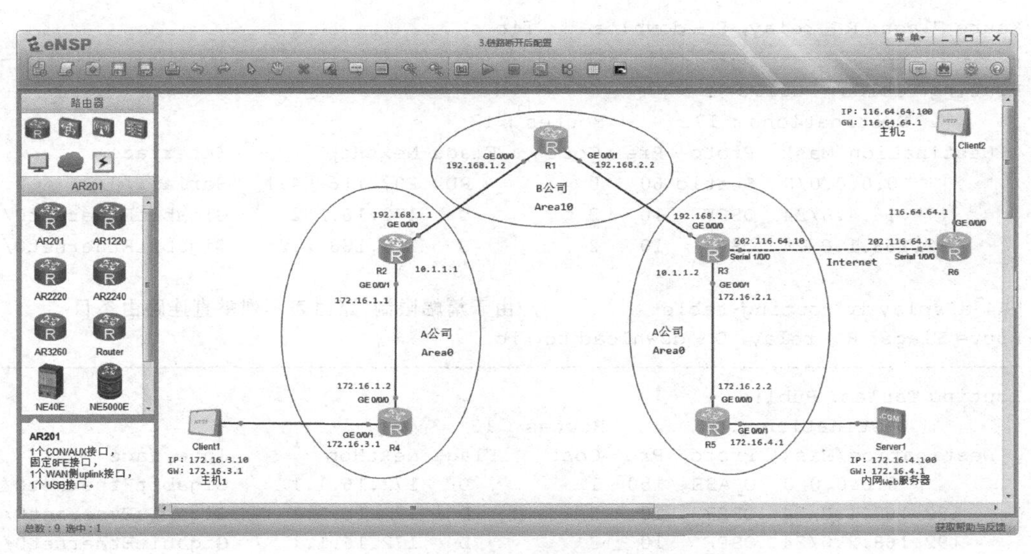

图 8-6 骨干区域被分割成两个逻辑区域

由于路由器 R1 为非骨干区域边界路由器(连接 Area0 骨干区域和其他非骨干区域的路由器),不能通告区域间 OSPF 路由,导致 Area0 两个逻辑区域相互隔离,没有彼此之间路由信息。重新查看路由表如下:

```
[R1]display ip routing-table          //由于篇幅限制,路由表不列举直连路由条目
Route Flags: R - relay, D - download to fib
------------------------------------------------------------------------------
Routing Tables: Public
         Destinations : 15          Routes : 15
  Destination/Mask  Proto  Pre  Cost     Flags  NextHop        Interface
        0.0.0.0/0   O_ASE  150  1          D    192.168.2.1    GigabitEthernet0/0/1
     172.16.1.0/24  OSPF   10   2          D    192.168.1.1    GigabitEthernet0/0/0
     172.16.2.0/24  OSPF   10   2          D    192.168.2.1    GigabitEthernet0/0/1
     172.16.3.0/24  OSPF   10   3          D    192.168.1.1    GigabitEthernet0/0/0
     172.16.4.0/24  OSPF   10   3          D    192.168.2.1    GigabitEthernet0/0/1
```

可以看到:

① 将<10.1.1.0>网段断开后,两边接口 G2/0/0 接口处于 Down 状态,OSPF 将不再通告其直连网段信息,即路由器 R2 和路由器 R3 的 network 10.1.1.0 0.0.0.255 命令自然失效。

② 路由器 R1 有去往 A 公司两个逻辑区域路由信息,但是不能通告区域间路由条目。

```
[R2]display ip routing-table          //由于篇幅限制,路由表不列举直连路由条目
Route Flags: R - relay, D - download to fib
------------------------------------------------------------------------------
Routing Tables: Public
         Destinations : 13          Routes : 13
  Destination/Mask  Proto  Pre  Cost     Flags  NextHop        Interface
        0.0.0.0/0   O_ASE  150  1          D    192.168.1.2    GigabitEthernet0/0/0
     172.16.3.0/24  OSPF   10   2          D    172.16.1.2     GigabitEthernet0/0/1
    192.168.2.0/24  OSPF   10   2          D    192.168.1.2    GigabitEthernet0/0/0

[R3]display ip routing-table          //由于篇幅限制,路由表不列举直连路由条目
```

```
Route Flags: R - relay, D - download to fib
------------------------------------------------------------------------------
Routing Tables: Public
        Destinations : 17        Routes : 17
  Destination/Mask   Proto  Pre  Cost      Flags NextHop         Interface
        0.0.0.0/0    Static 60   0           RD  202.116.64.1    Serial1/0/0
    172.16.4.0/24    OSPF   10   2           D   172.16.2.2      GigabitEthernet0/0/1
   192.168.1.0/24    OSPF   10   2           D   192.168.2.2     GigabitEthernet0/0/0

[R4]display ip routing-table          //由于篇幅限制，路由表不列举直连路由条目
Route Flags: R - relay, D - download to fib
------------------------------------------------------------------------------
Routing Tables: Public
        Destinations : 13        Routes : 13
  Destination/Mask   Proto  Pre  Cost      Flags NextHop         Interface
        0.0.0.0/0    O_ASE  150  1           D   172.16.1.1      GigabitEthernet0/0/0
   192.168.1.0/24    OSPF   10   2           D   172.16.1.1      GigabitEthernet0/0/0
   192.168.2.0/24    OSPF   10   3           D   172.16.1.1      GigabitEthernet0/0/0

[R5]display ip routing-table          //由于篇幅限制，路由表不列举直连路由条目
Route Flags: R - relay, D - download to fib
------------------------------------------------------------------------------
Routing Tables: Public
        Destinations : 13        Routes : 13
  Destination/Mask   Proto  Pre  Cost      Flags NextHop         Interface
        0.0.0.0/0    O_ASE  150  1           D   172.16.2.1      GigabitEthernet0/0/0
   192.168.1.0/24    OSPF   10   3           D   172.16.2.1      GigabitEthernet0/0/0
   192.168.2.0/24    OSPF   10   2           D   172.16.2.1      GigabitEthernet0/0/0
```

由路由器R2～R5可以看到，Area0逻辑区域之间缺少彼此OSPF路由条目，区域之间无法连通。路由器R1虽然不能转发区域间路由信息，但仍可转发区域间OSPF默认路由通告。

(7) 配置OSPF虚链路实现Area0骨干区域逻辑连接。

```
[R2]ospf 1 router-id 192.168.1.1              //在OSPF进程1中定义路由器router-id值。虽
                                                然路由器router-id可以通过选举自动产生，但
                                                不想花费时间查，手动重新定义即可
[R2-ospf-1]area 10
[R2-ospf-1-area-0.0.0.10]vlink-peer ?
  IP_ADDR<X.X.X.X> Neighbor router ID          //vlink-peer后面接router ID，router
                                                 ID值是IP地址格式
[R2-ospf-1-area-0.0.0.10]vlink-peer 192.168.2.1
[R2-ospf-1-area-0.0.0.10]quit
[R2-ospf-1]quit
[R2]
------------------------------------------------------------------------------
[R3]ospf 1 router-id 192.168.2.1
[R3-ospf-1]area 10
[R3-ospf-1-area-0.0.0.10]vlink-peer 192.168.1.1
[R3-ospf-1-area-0.0.0.10]quit
[R3-ospf-1]quit
[R3]
```

注意：vlink-peer指定router ID的IP地址，链路之间必须要能连通，否则虚链路无法连

接。即逻辑链路要能连通，前提是物理链路要能连通。以下是错误配置。

```
[R2]ospf 1 router-id 10.1.1.1
[R2-ospf-1]area 10
[R2-ospf-1-area-0.0.0.10]vlink-peer 10.1.1.2
[R2-ospf-1-area-0.0.0.10]quit
[R2-ospf-1]quit
[R2]
----------------------------------------------------------------------------------------
[R3-ospf-1]ospf 1 router-id 10.1.1.2
[R3-ospf-1]area 10
[R3-ospf-1-area-0.0.0.10]vlink-peer 10.1.1.1
[R3-ospf-1-area-0.0.0.10]quit
[R3-ospf-1]quit
[R3]
```

由于 IP 地址 10.1.1.1 与 10.1.1.2 之间无法连通（双方的 G2/0/0 物理接口处于 Down 状态，IP 自然失效），无法建立虚链路。

2. 任务验证

(1) 查看虚链路状态。

```
[R2]display ospf vlink
      OSPF Process 1 with Router ID 192.168.1.1
          Virtual Links
 Virtual-link Neighbor-id -> 192.168.2.1, Neighbor-State: Full
 Interface: 192.168.1.1 (GigabitEthernet0/0/0)
 Cost: 2 State: P-2-P Type: Virtual
 Transit Area: 0.0.0.10
 Timers: Hello 10 , Dead 40 , Retransmit 5 , Transmit Delay 1
 GR State: Normal
```

可以看到，Router ID 192.168.1.1(R2)已成功与 Router ID 192.168.2.1(R3)建立 OSPF 虚链路，虚链路状态为 Full。

(2) 重新查看路由器 R4 路由表。

```
[R4]display ip routing-table         //由于篇幅限制，路由表不列举直连路由条目
Route Flags: R - relay, D - download to fib
----------------------------------------------------------------------------------------
Routing Tables: Public
         Destinations : 15        Routes : 15
  Destination/Mask  Proto  Pre  Cost     Flags NextHop        Interface
         0.0.0.0/0  O_ASE  150  1           D  172.16.1.1     GigabitEthernet0/0/0
     172.16.2.0/24  OSPF   10   4           D  172.16.1.1     GigabitEthernet0/0/0
     172.16.4.0/24  OSPF   10   5           D  172.16.1.1     GigabitEthernet0/0/0
    192.168.1.0/24  OSPF   10   2           D  172.16.1.1     GigabitEthernet0/0/0
    192.168.2.0/24  OSPF   10   3           D  172.16.1.1     GigabitEthernet0/0/0
```

可以看到，除<10.1.1.0>网段外。路由器 R4 拥有公司内网全部路由信息。

(3) 连通性测试。

以路由器 R4 为例，R4 可以连通内网服务器和公网主机 2。

```
[R4]ping 116.64.64.100
  PING 116.64.64.100: 56 data bytes, press CTRL_C to break
    Reply from 116.64.64.100: bytes=56 Sequence=1 ttl=251 time=30 ms
    Reply from 116.64.64.100: bytes=56 Sequence=2 ttl=251 time=30 ms
    Reply from 116.64.64.100: bytes=56 Sequence=3 ttl=251 time=30 ms
  --- 116.64.64.100 ping statistics ---
    3 packet(s) transmitted
    3 packet(s) received
    0.00% packet loss
    round-trip min/avg/max = 30/30/30 ms

[R4]ping 172.16.4.100
  PING 172.16.4.100: 56 data bytes, press CTRL_C to break
    Reply from 172.16.4.100: bytes=56 Sequence=1 ttl=251 time=40 ms
    Reply from 172.16.4.100: bytes=56 Sequence=2 ttl=251 time=30 ms
    Reply from 172.16.4.100: bytes=56 Sequence=3 ttl=251 time=30 ms
  --- 172.16.4.100 ping statistics ---
    3 packet(s) transmitted
    3 packet(s) received
    0.00% packet loss
    round-trip min/avg/max = 30/33/40 ms
```

【任务总结】

(1) vlink-peer 指定 Router ID 的 IP 地址，之间必须要能连通，否则虚链路无法连接。

(2) 虚链路两端的 IP，可以处于同一网段，也可以处于不用网段，只要能够连通即可。

(3) 由于虚链路会穿越其他区域，一般建议使用 OSPF 区域认证提高安全性。

```
[R2]ospf 1
[R2-ospf-1]area 10
[R2-ospf-1-area-0.0.0.10]vlink-peer 192.168.2.1 hmac-md5 1 cipher gdcp
[R2-ospf-1-area-0.0.0.10]quit
[R2-ospf-1]quit
[R3]
------------------------------------------------------------------------------
[R3]ospf 1
[R3-ospf-1]area 10
[R3-ospf-1-area-0.0.0.10]vlink-peer 192.168.1.1 hmac-md5 1 cipher gdcp
[R3-ospf-1-area-0.0.0.10]quit
[R3-ospf-1]quit
[R3]
```

(4) 非骨干区域边界路由器不能通告区域间路由条目，但仍可转发区域间 OSPF 默认路由通告。

工作任务九

路由引入

【工作目的】

掌握如何在 RIP 和 OSPF 路由协议中相互引入，并修改开销值。

【工作任务】

A 公司＜172.16.0.0＞运行 RIPv2 协议，B 公司＜192.168.0.0＞运行 OSPF 协议。A 公司收购 B 公司后，不想对现有拓扑和 IP 规划进行大规模改动，打算新增路由器 R9 将两公司相连并连接至公网，并把服务器群统一迁至＜10.3.3.0＞网段。因此需要管理员在路由器 R9 上配置路由引入，实现两公司不同路由协议之间的通告，具体任务如下。

(1) 在 A 公司路由器上配置 RIPv2 协议，在 B 公司路由器上配置 OSPF 协议。

(2) 在路由器 R9 上同时配置 RIPv2 和 OSPF 协议，并相互引入。

(3) 主机 1(A 公司)和主机 2(B 公司)能够访问内网服务器群，也能够访问公网服务器。

【工作背景】

大型网络可能同时使用多种动态路由协议(如存在多种厂商设备，企业网络合并等)，但不同路由协议之间无法相互学习路由信息，导致局部网络之间无法连通。为解决这一问题，思科采用路由重发布(route redistribution)技术，将某种路由协议转换并封装成另外一种协议进行发布；华为称之为路由引入(route importation)，如将 A 路由协议引入 B 路由协议，从而 B 协议可以自动获得 A 协议的所有路由信息。

【任务分析】

1. 路由引入与开销

路由引入是指将路由信息从一种路由协议发布到另一种路由协议的操作。通过路由引入，可以实现路由信息在不同路由协议间传递。

OSPF 引入外部路由时，默认采用 Type 2 的方式，即不计算外部开销；但是引入如使用 Type 1 时则需要计算路由域外部开销。

OSPF 中引入外部 RIP 协议时，默认内部开销为 1，外部开销默认 Type 2 计量方式为 0，即总路由成本为 1+0=1；RIP 中引入其他的外部协议时，内部开销为实际跳数值，默认外部开销为 0，即总路由成本＝抵达边界路由器跳数+0。

2. 路由协议默认优先级

华为路由器分别定义了外部优先级和内部优先级。内部优先级参数值禁止用户修改，见

表 9-1。外部优先级用户可以手动修改具体参数,默认情况下外部优先级参数见表 9-2。

表 9-1 路由协议内部优先级

路由协议的类型	注 析	路由协议的内部优先级
Direct		0
OSPF		10
IS-IS Level-1		15
IS-IS Level-2		18
Static		60
RIP		100
OSPF ASE	引入外部路由协议的路由信息	150
OSPF NSSA	能学习本区域连接的外部路由,不能学习其他区域的外部路由	150
IBGP		200
EBGP		20

表 9-2 路由协议默认时的外部优先级

路由协议的类型	注 析	路由协议的内部优先级
Direct		0
OSPF		10
IS-IS		15
Static		60
RIP		100
OSPF ASE	引入外部路由协议的路由信息	150
OSPF NSSA	能学习本区域连接的外部路由,不能学习其他区域的外部路由	150
IBGP	IBGP 引入的外部路由	255
EBGP	EBGP 引入的外部路由	255

【设备器材】

路由器(AR1220)10 台,需要添加 1GEC 千兆接口模块或 2SA 串口模块。

主机 4 台,承担角色见表 9-3。

表 9-3 主机配置表

角 色	接 入 方 式	IP 地址	所 属 公 司
主机 1	eNSP PC 接入	172.16.10.10/24	公司 A
主机 2	eNSP PC 接入	192.168.10.10/24	公司 B
公网服务器	eNSP PC 接入	116.64.64.100/24	
私网服务器	eNSP PC 接入	10.3.3.10/24	收购后总公司

【环境拓扑】

环境拓扑如图 9-1 所示。

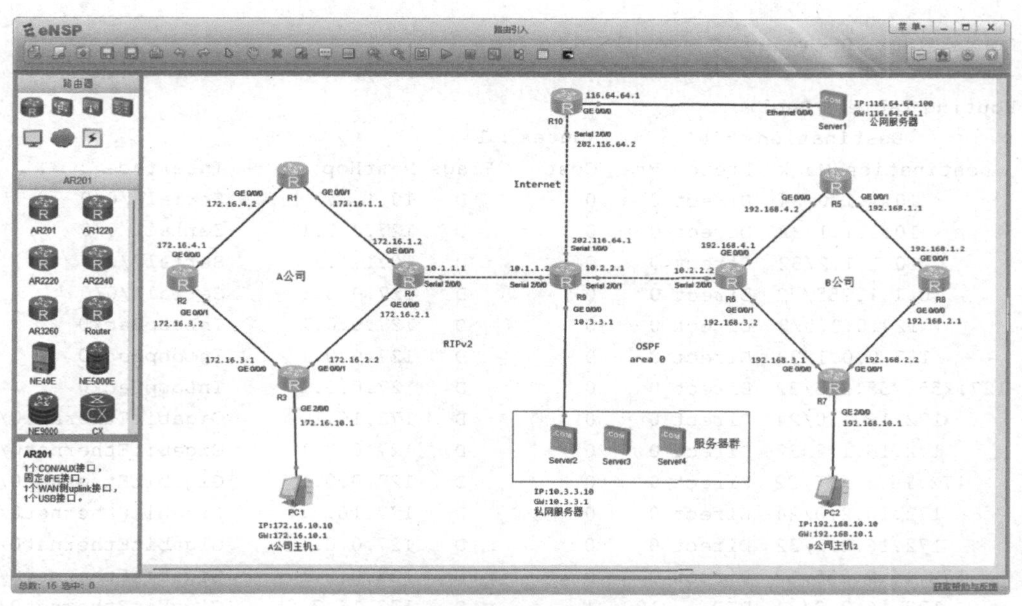

图 9-1　环境拓扑

【工作过程】

1. 基本配置

(1) 路由器接口 IP 和系统名称配置。

请读者根据网络拓扑自行配置路由器接口 IP 和系统名称，注意所有接口子网掩码长度均为 24。

(2) 在 A 公司配置 RIPv2 路由协议并关闭路由汇总。

```
[R4]rip 1
[R4-rip-1]version 2
[R4-rip-1]undo summary
[R4-rip-1]network 172.16.0.0
[R4-rip-1]network 10.0.0.0
[R4-rip-1]quit
[R4]
--------------------------------------------------------------------------------
[R3]rip 1
[R3-rip-1]version 2
[R3-rip-1]undo summary
[R3-rip-1]network 172.16.0.0
[R3-rip-1]quit
[R3]
```

请读者继续在路由器 R1 和路由器 R2 上配置 RIPv2 路由协议，以主类宣告方式(请勿宣告到子网)宣告直连网段。由于关闭路由汇总，配置完成后在路由器 R4 上可以看到 A 公司所有子网信息。

在路由器 R4 上查看路由表。

```
[R4]display ip routing-table
```

```
Route Flags: R - relay, D - download to fib
------------------------------------------------------------------------------
Routing Tables: Public
         Destinations : 17        Routes : 17
  Destination/Mask  Proto  Pre  Cost     Flags NextHop         Interface
        10.1.1.0/24  Direct 0    0          D   10.1.1.1        Serial2/0/0
        10.1.1.1/32  Direct 0    0          D   127.0.0.1       Serial2/0/0
        10.1.1.2/32  Direct 0    0          D   10.1.1.2        Serial2/0/0
      10.1.1.255/32  Direct 0    0          D   127.0.0.1       Serial2/0/0
        127.0.0.0/8  Direct 0    0          D   127.0.0.1       InLoopBack0
       127.0.0.1/32  Direct 0    0          D   127.0.0.1       InLoopBack0
 127.255.255.255/32  Direct 0    0          D   127.0.0.1       InLoopBack0
      172.16.1.0/24  Direct 0    0          D   172.16.1.2      GigabitEthernet0/0/1
      172.16.1.2/32  Direct 0    0          D   127.0.0.1       GigabitEthernet0/0/1
    172.16.1.255/32  Direct 0    0          D   127.0.0.1       GigabitEthernet0/0/1
      172.16.2.0/24  Direct 0    0          D   172.16.2.1      GigabitEthernet0/0/0
      172.16.2.1/32  Direct 0    0          D   127.0.0.1       GigabitEthernet0/0/0
    172.16.2.255/32  Direct 0    0          D   127.0.0.1       GigabitEthernet0/0/0
      172.16.3.0/24  RIP    100  1          D   172.16.2.2      GigabitEthernet0/0/0
      172.16.4.0/24  RIP    100  1          D   172.16.1.1      GigabitEthernet0/0/1
     172.16.10.0/24  RIP    100  1          D   172.16.2.2      GigabitEthernet0/0/0
 255.255.255.255/32  Direct 0    0          D   127.0.0.1       InLoopBack0
```

(3) 在 B 公司配置 OSPF 路由协议。

```
[R6]ospf 1
[R6-ospf-1]area 0
[R6-ospf-1-area-0.0.0.0]network 10.2.2.0 0.0.0.255
[R6-ospf-1-area-0.0.0.0]network 192.168.3.0 0.0.0.255
[R6-ospf-1-area-0.0.0.0]network 192.168.4.0 0.0.0.255
[R6-ospf-1]quit
[R6]
```

请读者继续在路由器 R5、R7 和 R8 上配置 OSPF 路由协议，并宣告所有接口直连网段，配置完成后在路由器 R6 上可以看到 B 公司所有网段信息。

```
[R6]display ip routing-table
Route Flags: R - relay, D - download to fib
------------------------------------------------------------------------------
Routing Tables: Public
         Destinations : 17        Routes : 17
  Destination/Mask  Proto  Pre  Cost     Flags NextHop         Interface
        10.2.2.0/24  Direct 0    0          D   10.2.2.2        Serial2/0/0
        10.2.2.1/32  Direct 0    0          D   10.2.2.1        Serial2/0/0
        10.2.2.2/32  Direct 0    0          D   127.0.0.1       Serial2/0/0
      10.2.2.255/32  Direct 0    0          D   127.0.0.1       Serial2/0/0
        127.0.0.0/8  Direct 0    0          D   127.0.0.1       InLoopBack0
       127.0.0.1/32  Direct 0    0          D   127.0.0.1       InLoopBack0
 127.255.255.255/32  Direct 0    0          D   127.0.0.1       InLoopBack0
     192.168.1.0/24  OSPF   10   2          D   192.168.4.2     GigabitEthernet0/0/0
     192.168.2.0/24  OSPF   10   2          D   192.168.3.1     GigabitEthernet0/0/1
     192.168.3.0/24  Direct 0    0          D   192.168.3.2     GigabitEthernet0/0/1
```

```
      192.168.3.2/32  Direct 0    0        D   127.0.0.1       GigabitEthernet0/0/1
    192.168.3.255/32  Direct 0    0        D   127.0.0.1       GigabitEthernet0/0/1
      192.168.4.0/24  Direct 0    0        D   192.168.4.1     GigabitEthernet0/0/0
      192.168.4.1/32  Direct 0    0        D   127.0.0.1       GigabitEthernet0/0/0
    192.168.4.255/32  Direct 0    0        D   127.0.0.1       GigabitEthernet0/0/0
     192.168.10.0/24  OSPF   10   2        D   192.168.3.1     GigabitEthernet0/0/1
  255.255.255.255/32  Direct 0    0        D   127.0.0.1       InLoopBack0
```

(4) 在路由器 R9 上配置双路由协议和双向路由引入、默认路由和 Easy-IP。

① 配置双路由协议和双向路由引入。

```
[R9]rip 1
[R9-rip-1]version 2
[R9-rip-1]undo summary
[R9-rip-1]network 10.0.0.0                //向A公司路由同时宣告<10.1.1.0>、<10.2.2.0>
                                            和<10.3.3.0>三个网段
[R9-rip-1]import-route ospf 1             //在 rip 进程 1 中引入 ospf 进程 1 的 路由条目
[R9-rip-1]quit
[R9]ospf 1
[R9-ospf-1]area 0
[R9-ospf-1-area-0.0.0.0]network 10.2.2.0 0.0.0.255
[R9-ospf-1-area-0.0.0.0]network 10.3.3.0 0.0.0.255
[R9-ospf-1-area-0.0.0.0]quit
[R9-ospf-1]import-route rip 1             //在 ospf 进程 1 中引入 rip 进程 1 的路由条目
[R9-ospf-1]quit
[R9]
```

② 配置 Easy-IP。

```
[R9]acl 2000
[R9-acl-basic-2000]rule permit source any
[R9-acl-basic-2000]quit
[R9]interface Serial 1/0/0
[R9-Serial1/0/0]nat outbound 2000
[R9-Serial1/0/0]quit
[R9]
```

③ 配置默认路由。

```
[R9]ip route-static 0.0.0.0 0.0.0.0 202.116.64.2
```

(5) 内外网连通性测试。

① 在路由器 R3 上查看路由表。由于篇幅限制,路由表不列举直连路由条目。

```
[R3]display ip routing-table
Route Flags: R - relay, D - download to fib
------------------------------------------------------------------------------
Routing Tables: Public
         Destinations : 23        Routes : 23
  Destination/Mask    Proto   Pre  Cost      Flags NextHop         Interface
        10.1.1.0/24   RIP     100  1           D   172.16.2.1      GigabitEthernet0/0/1
        10.2.2.0/24   RIP     100  2           D   172.16.2.1      GigabitEthernet0/0/1
        10.3.3.0/24   RIP     100  2           D   172.16.2.1      GigabitEthernet0/0/1
```

```
        172.16.1.0/24  RIP    100   1           D  172.16.2.1      GigabitEthernet0/0/1
        172.16.4.0/24  RIP    100   1           D  172.16.3.2      GigabitEthernet0/0/0
       192.168.1.0/24  RIP    100   2           D  172.16.2.1      GigabitEthernet0/0/1
       192.168.2.0/24  RIP    100   2           D  172.16.2.1      GigabitEthernet0/0/1
       192.168.3.0/24  RIP    100   2           D  172.16.2.1      GigabitEthernet0/0/1
       192.168.4.0/24  RIP    100   2           D  172.16.2.1      GigabitEthernet0/0/1
      192.168.10.0/24  RIP    100   2           D  172.16.2.1      GigabitEthernet0/0/1
```

可以看到，路由器R9将B公司ospf路由条目引入rip 1后，在路由器R3上可以发现B公司所有网段。由于在RIP中引入其他的外部协议时默认内部开销为0，因此去往B公司所有网段开销值均为2(去往路由器R9的开销值2＋外部开销值0)，见上述黑体字。

② 在R7查看路由表。

由于篇幅限制，路由表不列举直连路由条目。

```
[R7]display ip routing-table
Route Flags: R - relay, D - download to fib
------------------------------------------------------------------------------
Routing Tables: Public
         Destinations : 27       Routes : 27
  Destination/Mask    Proto  Pre  Cost      Flags NextHop         Interface
        10.1.1.0/24   O_ASE  150  1           D   192.168.3.2     GigabitEthernet0/0/0
        10.2.2.0/24   OSPF   10   49          D   192.168.3.2     GigabitEthernet0/0/0
        10.3.3.0/24   OSPF   10   50          D   192.168.3.2     GigabitEthernet0/0/0
      172.16.1.0/24   O_ASE  150  1           D   192.168.3.2     GigabitEthernet0/0/0
      172.16.2.0/24   O_ASE  150  1           D   192.168.3.2     GigabitEthernet0/0/0
      172.16.3.0/24   O_ASE  150  1           D   192.168.3.2     GigabitEthernet0/0/0
      172.16.4.0/24   O_ASE  150  1           D   192.168.3.2     GigabitEthernet0/0/0
     172.16.10.0/24   O_ASE  150  1           D   192.168.3.2     GigabitEthernet0/0/0
     192.168.1.0/24   OSPF   10   2           D   192.168.2.1     GigabitEthernet0/0/1
     192.168.4.0/24   OSPF   10   2           D   192.168.3.2     GigabitEthernet0/0/0
```

可以看到，路由器R9将A公司rip路由条目引入ospf 1后，在路由器R7上可以发现A公司所有网段。其中O_ASE表示OSPF引入的外部路由协议，优先级为150。去往<10.2.2.0>网段开销值为49(1＋48)，去往<10.3.3.0>开销值为50(1＋48＋1)。OSPF引入外部路由时，默认采用Type 2的方式，即不计算外部的开销，因此去往A公司<172.16.0.0>网段开销值均为1(OSPF中引入外部RIP协议时，默认的内部开销为1＋外部开销0)，去往<10.1.1.0>网段开销值为1(1＋0)。

③ 外网连通性测试。

从上述公司A和公司B路由表中可以看到，路由器R9通告时仅引入RIP/OSPF协议路由，未引入公网路由，因此公司A和公司B之间可以互通，但是不能连接外网。以路由器R7为例，与公网连通性测试如下：

```
[R7]ping 202.116.64.1
  PING 202.116.64.1: 56 data bytes, press CTRL_C to break
    Request time out
    Request time out
    Request time out
    Request time out
    Request time out
```

```
--- 202.116.64.1 ping statistics ---
  5 packet(s) transmitted
  0 packet(s) received
  100.00% packet loss
```

(6) 在路由器 R9 中通告默认路由。

```
[R9]rip 1
[R9-rip-1]default-route originate          //在 RIP 中下发默认路由通告
[R9-rip-1]quit
[R9]ospf 1
[R9-ospf-1]default-route-advertise         //在 OSPF 中下发默认路由通告
[R9-ospf-1]quit
[R9]
```

注意：

① [R9-ospf-1]default-route-advertise default-route-advertise：向其他路由器通告本地路由器 R9 是它们的默认路由，前提是路由器 R9 必须存在默认路由。假如路由器 R9 没有配置默认路由，则不向其他路由器通告下发默认路由。

② [R9-ospf-1]default-route-advertise default-route-advertise always：无论本地路由器 R9 是否配置默认路由，都向其他路由器通告下发默认路由。

此时在路由器 R3 上查看路由表如下。由于篇幅限制，路由表不列举直连路由条目。

```
[R3]display ip routing-table
Route Flags: R - relay, D - download to fib
------------------------------------------------------------------------------
Routing Tables: Public
         Destinations : 24        Routes : 24
  Destination/Mask  Proto  Pre  Cost     Flags NextHop         Interface
         0.0.0.0/0  RIP    100  2          D   172.16.2.1      GigabitEthernet0/0/1
       10.1.1.0/24  RIP    100  1          D   172.16.2.1      GigabitEthernet0/0/1
       10.2.2.0/24  RIP    100  2          D   172.16.2.1      GigabitEthernet0/0/1
       10.3.3.0/24  RIP    100  2          D   172.16.2.1      GigabitEthernet0/0/1
     172.16.1.0/24  RIP    100  1          D   172.16.2.1      GigabitEthernet0/0/1
     172.16.4.0/24  RIP    100  1          D   172.16.3.2      GigabitEthernet0/0/0
    192.168.1.0/24  RIP    100  2          D   172.16.2.1      GigabitEthernet0/0/1
    192.168.2.0/24  RIP    100  2          D   172.16.2.1      GigabitEthernet0/0/1
    192.168.3.0/24  RIP    100  2          D   172.16.2.1      GigabitEthernet0/0/1
    192.168.4.0/24  RIP    100  2          D   172.16.2.1      GigabitEthernet0/0/1
   192.168.10.0/24  RIP    100  2          D   172.16.2.1      GigabitEthernet0/0/1
```

可以看到，通过管理员指定通告默认路由，在路由器 R3 路由表中才会出现 0.0.0.0 RIP 协议类型的默认路由。

在路由器 R7 上查看路由表如下。由于篇幅限制，路由表不列举直连路由条目。

```
[R7]display ip routing-table
Route Flags: R - relay, D - download to fib
------------------------------------------------------------------------------
Routing Tables: Public
         Destinations : 24        Routes : 24
  Destination/Mask  Proto  Pre  Cost     Flags NextHop         Interface
```

```
    0.0.0.0/0  O_ASE  150  1        D  192.168.3.2   GigabitEthernet0/0/0
   10.1.1.0/24  O_ASE  150  1        D  192.168.3.2   GigabitEthernet0/0/0
   10.2.2.0/24  OSPF   10   49       D  192.168.3.2   GigabitEthernet0/0/0
   10.3.3.0/24  OSPF   10   50       D  192.168.3.2   GigabitEthernet0/0/0
  172.16.1.0/24  O_ASE  150  1        D  192.168.3.2   GigabitEthernet0/0/0
  172.16.2.0/24  O_ASE  150  1        D  192.168.3.2   GigabitEthernet0/0/0
  172.16.3.0/24  O_ASE  150  1        D  192.168.3.2   GigabitEthernet0/0/0
  172.16.4.0/24  O_ASE  150  1        D  192.168.3.2   GigabitEthernet0/0/0
 172.16.10.0/24  O_ASE  150  1        D  192.168.3.2   GigabitEthernet0/0/0
 192.168.1.0/24  OSPF   10   2        D  192.168.2.1   GigabitEthernet0/0/1
 192.168.4.0/24  OSPF   10   2        D  192.168.3.2   GigabitEthernet0/0/0
```

同样,通过管理员指定通告默认路由,在路由器 R7 路由表中也出现 0.0.0.0 O_ASE (OSPF 引入的外部路由)类型的默认路由。

2. 任务验证

(1) 内网连通性验证。

主机 1 和主机 2 能够相互连通,通过主机 1 测试连通情况如图 9-2 所示。

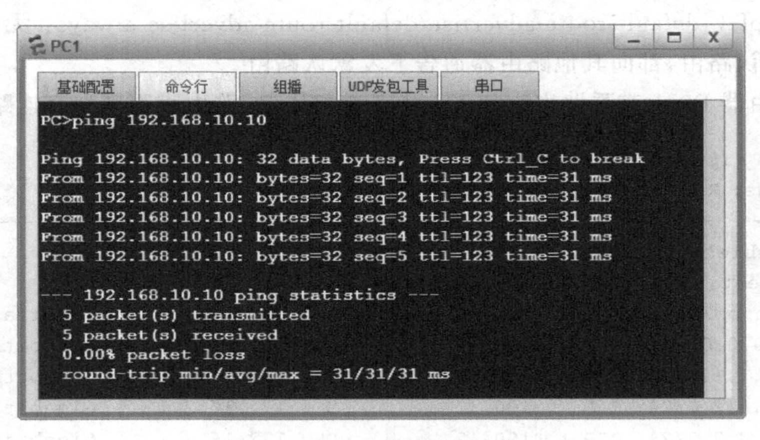

图 9-2 主机 1 能够连通主机 2

(2) 公网连通性验证。

主机 1 和主机 2 能够连通公网服务器。其中主机 1 连通情况如图 9-3 所示。

图 9-3 主机 1 能够连通公网服务器

【任务总结】

(1) 路由引入的操作,是在路由选择域的边界设备上完成,将路由信息从一个路由协议引入到另一个路由协议。

(2) 将路由信息从路由协议 A 引入路由协议 B,则是在路由协议 B 的配置视图下完成相关配置。

(3) 只有存在于路由表中的路由才能够被顺利引入。

工作任务十

IS-IS 路由协议

【工作目的】

掌握 IS-IS 路由协议配置的方法，理解 Level-1 类型的路由和网络实体标识方法。

【工作任务】

A 公司(172.16.0.0)收购 B 公司(192.168.0.0)，两公司通过路由器 R9 相连并连接至公网，将服务器群统一迁至(10.3.3.0)网段。考虑公司合并后，部门会发生变化，路由器接口 IP 地址也可能跟着变更。为方便管理，减少配置复杂度和组播流量，公司决定将网络拓扑分为 3 个区域，采用 IS-IS 路由协议互联，具体任务如下。

(1) 在 A 公司路由器上配置 IS-IS 协议，隔离其他区域路由组播通告。

(2) 在 B 公司路由器上配置 IS-IS 协议，隔离其他区域路由组播通告。

(3) 在路由器 R9 上配置 IS-IS 协议，向 A 公司和 B 公司通告区域间路由条目。

(4) 主机 1(A 公司)和主机 2(B 公司)能够访问内网服务器群，也能够访问公网服务器。

【工作背景】

无论是 RIP 路由还是 OSPF 路由，接口 IP 地址变更后路由协议需要重新配置。IS-IS 路由可以很好地解决这个问题，其配置方便快捷，性能优于 OSPF，接口 IP 地址变更后不影响网络的连通性，常用于组建大规模网络，也是电信运营商普遍选用的内部网关协议。

【任务分析】

IS-IS(intermediate system to intermediate system，中间系统到中间系统)是一种内部网关协议，是电信运营商普遍采用的内部网关协议之一。IS-IS 路由协议类似 OSPF 路由协议，是链路状态路由协议，也是使用最短路径优先算法 SPF(Dijkstra 算法)，不同的是 IS-IS 路由协议增加了增强型的最短路径优先算法 I-SPF。而且在大型网络的实际应用中，IS-IS 的使用频率比 OSPF 高，因为 IS-IS 性能优于 OSPF。

1. IS-IS 三种类型路由器

```
[Huawei]isis 1
[Huawei-isis-1]is-level ?
  level-1   Level-1
  level-1-2 Level-1-2
  level-2   Level-2
```

IS-IS 路由协议将路由器分为三种类型的路由器，分别是 level-1 路由器、level-2 路由器和 level-1-2 路由器。

(1) level-1。区域内路由，与同一区域的 Level-1 和 Level-1-2 路由器形成邻居关系；level-1 路由器中路由表仅存放本地 IS-IS 区域路由，去往其他区域路由通过默认路由器抵达。

(2) level-2。区域间路由，与 Level-2(其他区域路由器)和 Level-1-2 路由器形成邻居关系；level-2 路由器路由表中应存在其他区域 IS-IS 路由条目。

(3) level-1-2。同时属于 Level-1 和 Level-2 的路由器称为 Level-1-2 路由器。level-1-2 路由器中路由表应存在其他区域 IS-IS 路由条目。

2. IS-IS 网络实体名称

```
[Huawei-isis-1]network-entity 49.0001.0000.0000.0001.00
```

IS-IS 路由协议使用网络服务访问点 NSAP 来定位网络资源的地址，它包括了 IS-IS 区域，路由器的 System ID(相当于 Router ID)，协议标识符为 SEL，表示方法是 Area Address＋System ID＋SEL。

IS-IS Area＝49.0001，System ID＝0000.0000.0001，NSEL＝00

其中，Area ID 则通常在全网内统一指定，49 地址为私有地址，类似私有 IP 地址；System ID 通常由 MAC 地址构成或由 IP 地址转换而来。

【设备器材】

路由器(AR1220)11 台，需要添加 1GEC 千兆接口模块或 2SA 串口模块。

主机 4 台，承担角色见表 10-1。

表 10-1 主机配置表

角　色	接入方式	IP 地址	所属公司
主机 1	eNSP PC 接入	172.16.10.10/24	公司 A
主机 2	eNSP PC 接入	192.168.10.10/24	公司 B
公网服务器	eNSP PC 接入	116.64.64.100/24	
私网服务器	eNSP PC 接入	10.3.3.10/24	收购后总公司

【环境拓扑】

环境拓扑如图 10-1 所示。

【工作过程】

1. 基本配置

(1) 路由器接口 IP 和系统名称配置。

请读者根据网络拓扑自行配置路由器接口 IP 地址和系统名称，注意所有接口子网掩码长度均为 24。

(2) 在 A 公司和 B 公司路由器配置 IS-IS 协议。

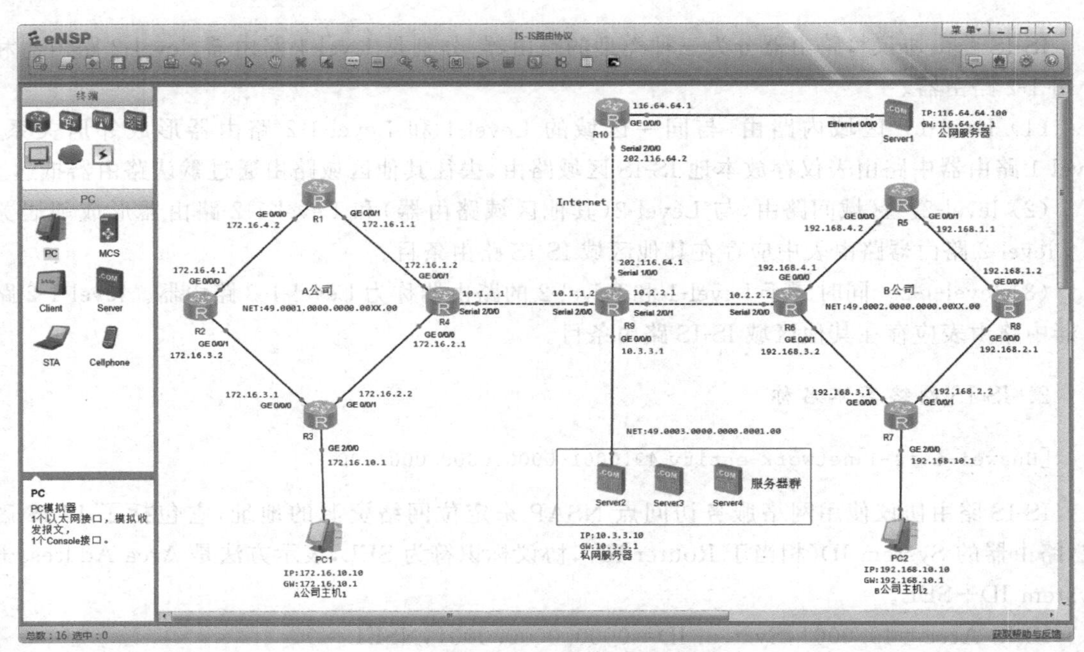

图 10-1 环境拓扑

在 A 公司配置 IS-IS 路由协议。

```
[R1]isis 1        //1 为 IS-IS 进程号,取值范围为 1～65535
[R1-isis-1]network-entity 49.0001.0000.0000.0001.00  //注意区域内网络实体名称不能冲
                                                      突,否则无法建立 ISIS 邻居
```

注:

① network-entity 用于配置 NET(network entity title,网络实体名称)。

② 必须配置 NET,否则 IS-IS 协议无法真正启动。

③ 网络实体名称由区域 ID+系统 ID+SEL 构成。本行命令区域 ID 是 49.0001(49 开头为私用区域),0000.0000.0001 是系统 ID(系统 ID 可以自定义,但一个区域内系统 ID 必须唯一,实际工程中推荐使用路由器 MAC 地址作为系统 ID),00 为 SEL(SEL 值必须为 00)。

④ 位于同一区域内的路由器 Area ID 必须相同,否则路由器无法通信。

```
[R1-isis-1]is-level level-1                //指定 R1 为 level-1 路由,只能学
                                            习区域内路由信息
[R1-isis-1]quit
[R1]interface GigabitEthernet 0/0/0
[R1-GigabitEthernet0/0/0]isis enable 1
[R1-GigabitEthernet0/0/0]quit
[R1]interface GigabitEthernet 0/0/1
[R1-GigabitEthernet0/0/1]isis enable 1
[R1-GigabitEthernet0/0/1]quit
[R1]
------------------------------------------------------------------------------
[R2]isis 1
[R2-isis-1]network-entity 49.0001.0000.0000.0010.00  //注意区域内网络实体名称不能冲
                                                      突,否则无法建立 ISIS 邻居
[R2-isis-1]is-level level-1
```

```
[R2-isis-1]quit
[R2]interface GigabitEthernet 0/0/0
[R2-GigabitEthernet0/0/0]isis enable 1
[R2-GigabitEthernet0/0/0]quit
[R2]interface GigabitEthernet 0/0/1
[R2-GigabitEthernet0/0/1]isis enable 1
[R2-GigabitEthernet0/0/1]quit
[R2]
-----------------------------------------------------------------------------------------
[R3]isis 1
[R3-isis-1]network-entity 49.0001.0000.0000.0011.00  //注意区域内网络实体名称不能冲
                                                        突,否则无法建立 ISIS 邻居
[R3-isis-1]is-level level-1
[R3-isis-1]quit
[R3]interface GigabitEthernet 0/0/0
[R3-GigabitEthernet0/0/0]isis enable 1
[R3-GigabitEthernet0/0/0]quit
[R3]interface GigabitEthernet 0/0/1
[R3-GigabitEthernet0/0/1]isis enable 1
[R3-GigabitEthernet0/0/1]quit
[R3]interface GigabitEthernet 2/0/0
[R3-GigabitEthernet2/0/0]isis enable 1
[R3-GigabitEthernet2/0/0]quit
[R3]
-----------------------------------------------------------------------------------------
[R4]isis 1
[R4-isis-1]network-entity 49.0001.0000.0000.0100.00  //注意区域内网络实体名称不能冲
                                                        突,否则无法建立 ISIS 邻居
[R4-isis-1]is-level level-1-2
[R4-isis-1]quit
[R4]interface GigabitEthernet 0/0/0
[R4-GigabitEthernet0/0/0]isis enable 1
[R4-GigabitEthernet0/0/0]quit
[R4]interface GigabitEthernet 0/0/1
[R4-GigabitEthernet0/0/1]isis enable 1
[R4-GigabitEthernet0/0/1]quit
[R4]interface Serial 2/0/0
[R4-Serial2/0/0]isis enable 1
[R4-Serial2/0/0]quit
[R4]
-----------------------------------------------------------------------------------------
```

按照上述步骤,请读者自行在 B 公司配置 IS-IS 路由协议。

(3) 在路由器 R9 上配置 IS-IS 协议和出口网关。

① 配置 IS-IS 协议。

```
[R9]isis 1
[R9-isis-1]network-entity 49.0003.0000.0000.0001.00  //49.0003 区域
[R9-isis-1]is-level level-2                          //指定 R9 为 level-2 路由,可以转
                                                       发区域间路由信息
[R9-isis-1]quit
[R9]interface Serial 2/0/0
[R9-Serial2/0/0]isis enable 1
```

```
[R9-Serial2/0/0]quit
[R9]interface Serial 2/0/1
[R9-Serial2/0/1]isis enable 1
[R9-Serial2/0/1]quit
[R9]interface GigabitEthernet 0/0/0
[R9-GigabitEthernet0/0/0]isis enable 1
[R9-GigabitEthernet0/0/0]quit
[R9]
```

② 配置 Easy-IP。

```
[R9]acl 2000
[R9-acl-basic-2000]rule permit source any
[R9-acl-basic-2000]quit
[R9]interface Serial 1/0/0
[R9-Serial1/0/0]nat outbound 2000
[R9-Serial1/0/0]quit
[R9]
```

③ 配置默认路由。

```
[R9]ip route-static 0.0.0.0 0.0.0.0 202.116.64.2
```

(4) 内外网连通性测试。

① 在 R2 查看路由表。

由于篇幅限制,路由表不列举直连路由条目。

```
[R2]display ip routing-table
Route Flags: R - relay, D - download to fib
------------------------------------------------------------------------------
Routing Tables: Public
         Destinations : 15        Routes : 17

  Destination/Mask    Proto   Pre  Cost      Flags NextHop         Interface
          0.0.0.0/0   ISIS-  L15   20          D   172.16.4.2      GigabitEthernet0/0/0
                      ISIS-  L15   20          D   172.16.3.1      GigabitEthernet0/0/1
        10.1.1.0/24   ISIS-  L15   30          D   172.16.4.2      GigabitEthernet0/0/0
                      ISIS-  L15   30          D   172.16.3.1      GigabitEthernet0/0/1
      172.16.1.0/24   ISIS-  L15   20          D   172.16.4.2      GigabitEthernet0/0/0
      172.16.2.0/24   ISIS-  L15   20          D   172.16.3.1      GigabitEthernet0/0/1
     172.16.10.0/24   ISIS-  L15   20          D   172.16.3.1      GigabitEthernet0/0/1
```

可以看到,R2 的 is-level 属于 level-1 类型,仅能看到 A 公司内网 IS-IS 网段信息,去往其他区域网段信息由 R4(is-level level-1-2)负责通告。

注意:当 level-1-2 路由器向 level-1 路由器通告区域间路由条目时,level-1-2 路由器会自动充当 level-1 路由器的网关,level-1 路由器会自动出现 ISIS-L1(0.0.0.0)型默认路由。

② 在路由器 R4 上查看路由表。

由于篇幅限制,路由表不列举直连路由条目。

```
[R4]display ip routing-table
Route Flags: R - relay, D - download to fib
------------------------------------------------------------------------------
```

```
Routing Tables: Public
         Destinations : 23        Routes : 23
  Destination/Mask  Proto    Pre  Cost     Flags NextHop        Interface
        10.2.2.0/24  ISIS- L2 15   20        D   10.1.1.2       Serial2/0/0
        10.3.3.0/24  ISIS- L2 15   20        D   10.1.1.2       Serial2/0/0
      172.16.3.0/24  ISIS- L1 15   20        D   172.16.2.2     GigabitEthernet0/0/0
      172.16.4.0/24  ISIS- L1 15   20        D   172.16.1.1     GigabitEthernet0/0/1
     172.16.10.0/24  ISIS- L1 15   20        D   172.16.2.2     GigabitEthernet0/0/0
     192.168.1.0/24  ISIS- L2 15   40        D   10.1.1.2       Serial2/0/0
     192.168.2.0/24  ISIS- L2 15   40        D   10.1.1.2       Serial2/0/0
     192.168.3.0/24  ISIS- L2 15   30        D   10.1.1.2       Serial2/0/0
     192.168.4.0/24  ISIS- L2 15   30        D   10.1.1.2       Serial2/0/0
    192.168.10.0/24  ISIS- L2 15   40        D   10.1.1.2       Serial2/0/0
```

可以看到，路由器 R4 属于 level-1-2 类型，能看到区域内路由和区域间路由。由于缺少默认路由，路由器 R4 无法连通 Internet。

③ 外网连通性测试。

```
[R4]ping 202.116.64.1
  PING 202.116.64.1: 56 data bytes, press CTRL_C to break
    Request time out
    Request time out
    Request time out
    Request time out
    Request time out
  --- 202.116.64.1 ping statistics ---
    5 packet(s) transmitted
    0 packet(s) received
    100.00% packet loss
```

注意：level-1-2 路由器一般充当网关，level-2 路由器作为区域间路由器，它们之间相互通告时不会自动产生 0.0.0.0 默认网段的 IS-IS 路由。

由以上路由表可以推断，此时内网能够相互连通，但是无法连接外网。其中主机 1 与内网服务器和主机 2 连通情况如图 10-2 所示。

(5) 在路由器 R9 中通告 IS-IS 默认路由。

```
[R9]isis 1
[R9-isis-1]default-route-advertise   //在 IS-IS 协议中通告默认路由，其他路由器通过 IS-IS
                                       协议学习到后，将下一跳指向发布默认路由的路由器
[R9-isis-1]quit
[R9]
```

注意：

① [R9-isis-1]default-route-advertise default-route-advertise：向其他路由器通告本地路由器 R9 是其默认路由，前提是路由器 R9 必须存在默认路由。假如路由器 R9 没有配置默认路由，则不向其他路由器通告下发默认路由。

② [R9-isis-1]default-route-advertise default-route-advertise always：无论本地路由器 R9 是否配置默认路由，都向其他路由器通告下发默认路由。

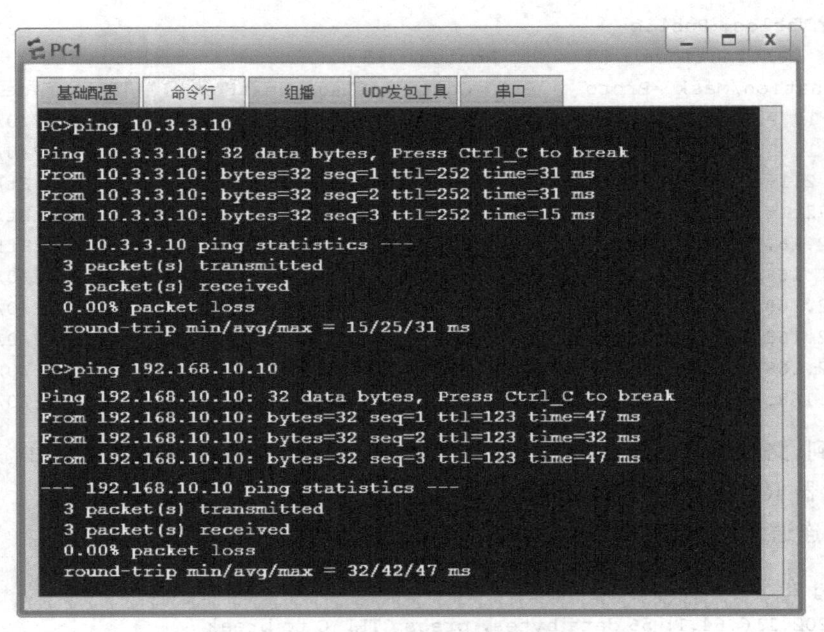

图 10-2　主机 1 能够连通内网服务器和主机 2

此时在路由器 R4 上查看路由表如下。由于篇幅限制，路由表不列举直连路由条目。

```
[R4]display ip routing-table
Route Flags: R - relay, D - download to fib
------------------------------------------------------------------------------
Routing Tables: Public
         Destinations : 24        Routes : 24

  Destination/Mask   Proto     Pre  Cost     Flags NextHop         Interface

        0.0.0.0/0    ISIS- L2  15   10         D   10.1.1.2        Serial2/0/0
       10.2.2.0/24   ISIS- L2  15   20         D   10.1.1.2        Serial2/0/0
       10.3.3.0/24   ISIS- L2  15   20         D   10.1.1.2        Serial2/0/0
     172.16.3.0/24   ISIS- L1  15   20         D   172.16.2.2      GigabitEthernet0/0/0
     172.16.4.0/24   ISIS- L1  15   20         D   172.16.1.1      GigabitEthernet0/0/1
    172.16.10.0/24   ISIS- L1  15   20         D   172.16.2.2      GigabitEthernet0/0/0
    192.168.1.0/24   ISIS- L2  15   40         D   10.1.1.2        Serial2/0/0
    192.168.2.0/24   ISIS- L2  15   40         D   10.1.1.2        Serial2/0/0
    192.168.3.0/24   ISIS- L2  15   30         D   10.1.1.2        Serial2/0/0
    192.168.4.0/24   ISIS- L2  15   30         D   10.1.1.2        Serial2/0/0
   192.168.10.0/24   ISIS- L2  15   40         D   10.1.1.2        Serial2/0/0
```

可以看到，通过管理员指定通告默认路由，路由器 R4 路由表中才会出现 ISIS-L2(0.0.0.0)型默认路由。

2. 任务验证

(1) 连通性验证。

主机 1 和主机 2 能够连通公网服务器。其中主机 1 连通情况如图 10-3 所示。

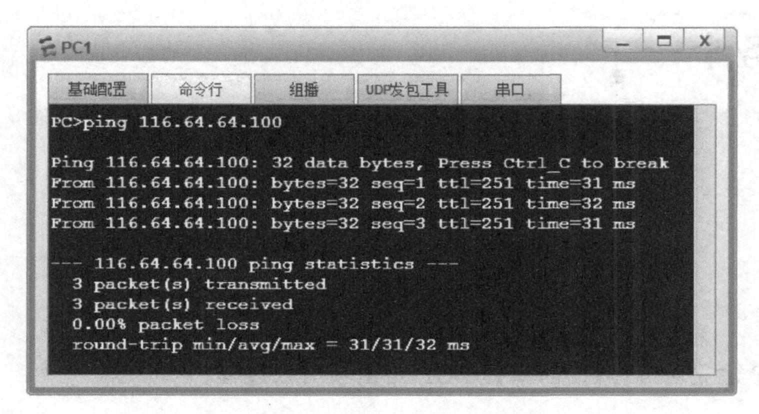

图 10-3 主机 1 能够连通公网服务器

(2) 变更公司 A 接口 IP 并进行 IS-IS 路由连通性验证(选做)。

将公司 A 路由器所有接口从(172.16.0.0)网段更改为(172.30.0.0),不影响网络连通性。可以看到,IS-IS 协议配置简单方便,当接口 IP 变更时路由协议不需重新配置(而 RIP 和 OSPF 需要重新宣告直连网段),是电信运营商和大型企业优先选用的内部网关协议。

【任务总结】

(1) level-1 类型路由器组成非骨干区域,level-2、level-1/2 类型路由器组成骨干区域。

(2) NET 网络实体标记名必须配置,除了同个区域的 Area ID 相同外,System ID 不能一样,并且 NEL 值是 00。

(3) IS-IS 路由协议不像其他的动态路由协议需要 network 路由器的网段信息来发布路由信息,而是在接口模式下启用 ISIS,便可宣告接口下网段信息。

(4) 默认情况下,IS-IS 路由器既是 level-1 型路由器,也是 level-2 型路由器,即默认情况下,IS-IS 路由器是 level-1-2 类型路由器。

(5) level-1 区域内型路由器仅存在所在区域路由条目,去往其他区域走默认路由(默认路由条目通过 level-1-2 路由器通告);而 level-1-2 和 level-2 型路由器存在区域间路由条目。

工作任务十一

EBGP 对等体与路由引入

【工作目的】

理解 EBGP 建立对等体过程，掌握 BGP 路由项引入过程。

【工作任务】

运营商路由器通过 BGP 协议互联，组成公网拓扑，为公司 A 和 Baidu 公司提供网络接入服务，具体任务如下。

(1) 路由器 R2、R3、R4 和 R5 分别隶属于 AS200、AS300、AS400 和 AS500，建立 EBGP 对等体以通告区域路由条目。

(2) 配置 Easy-IP，主机 1 和主机 2 可以访问公网。

(3) 配置 Nat Server，主机 1 和主机 2 可以访问 Baidu 公司 Web 服务和 FTP 服务。

【工作背景】

路由协议分为内部网关协议(interior gateway protocol，IGP)和外部网关协议(exterior gateway protocol，EGP)两类。AS 的内部使用 IGP 来计算和发现路由，其中常见的 IGP 包括 RIP、OSPF、IS-IS 等。不同 AS 之间的连接采用外部网关协议 EGP，而 EGP 只有 BGP(border gateway protocol)唯一协议。

BGP 虽然是一种动态路由协议，但它本身并不产生路由，也不发现路由，只是将 BGP 路由表转发给不同 AS 对等体，从而实现不同 AS 之间的互联，组成公网拓扑。

【任务分析】

BGP 属于外部网关协议，用于实现不同 AS 之间连接，采用 TCP 作为传输协议，端口号为 179。

1. IBGP 与 EBGP

IGP 协议可以自动发现邻居，不需人工干预。而在 BGP 中，需要手动建立对等体(邻居)。BGP 的邻居关系分为 IBGP(internal BGP)和 EBGP(external BGP)两种。

(1) IBGP。运行在相同 AS 内的 BGP 路由器建立的邻居关系为 IBGP 邻居关系(AS 编号相同)。

(2) EBGP。运行在不同 AS 之间的 BGP 路由器建立的邻居关系为 EBGP 邻居关系(AS 编号不同)。

2. BGP 路由引入

BGP 并不产生路由，也不发现路由，因此需要将其他路由(如 IGP 路由等)引入 BGP 路由表中，从而将这些路由在 AS 之内和 AS 之间传播。BGP 协议支持通过以下两种方式引入路由。

(1) Network。在 IP 路由表中将已经存在的路由项逐条引入 BGP 路由表，比 Import 方式更精确。

(2) Import。按协议类型，将 RIP 路由、OSPF 路由、ISIS 路由等协议的路由引入 BGP 路由表中。为了保证引入的 IGP 路由的有效性，Import 方式还可以引入静态路由和直连路由。

【设备器材】

三层交换机(S5700)2 台，路由器(AR1220)6 台，部分需添加 1GEC 千兆接口模块或 2SA 串口模块。

主机 4 台，承担角色见表 11-1。

表 11-1　主机配置表

角　　色	接 入 方 式	IP 地址	所 属 公 司
主机 1	eNSP PC 接入	172.16.10.10/24	公司 A 技术部
主机 2	eNSP PC 接入	172.168.20.10/24	公司 A 工程部
Web 服务器	eNSP PC 接入	192.168.10.10/24	Baidu 公司
FTP 服务器	eNSP PC 接入	192.168.20.10/24	Baidu 公司

【环境拓扑】

环境拓扑如图 11-1 所示。

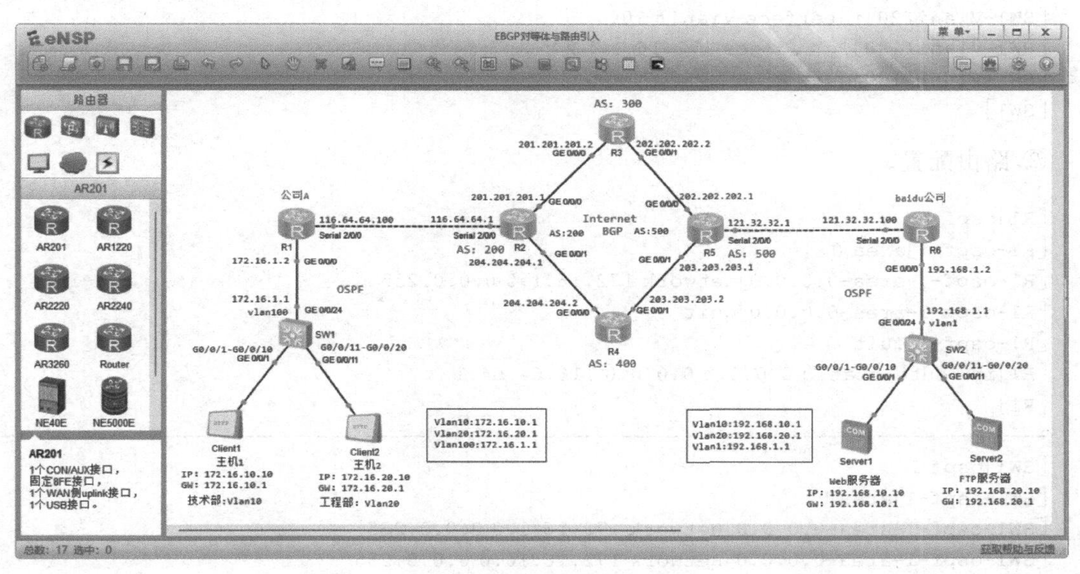

图 11-1　环境拓扑

【工作过程】

1. 基本配置

(1) 路由器接口 IP 和系统名称配置。

请读者根据网络拓扑自行配置路由器接口 IP 和系统名称，注意所有接口子网掩码长度均为 24。

(2) A 公司内网配置。

① Vlan 划分与 IP 配置。

```
[SW1]vlan batch 10 20 100
[SW1]port-group 1                                              //技术部组
[SW1-port-group-1]group-member GigabitEthernet 0/0/1 to GigabitEthernet 0/0/10
[SW1-port-group-1]port link-type access
[SW1-port-group-1]port default vlan 10
[SW1-port-group-1]quit
[SW1]port-group 2                                              //工程部组
[SW1-port-group-2]group-member GigabitEthernet 0/0/11 to GigabitEthernet 0/0/20
[SW1-port-group-2]port link-type access
[SW1-port-group-2]port default vlan 20
[SW1-port-group-2]quit
[SW1]interface GigabitEthernet 0/0/24
[SW1-GigabitEthernet0/0/24]port link-type trunk
[SW1-GigabitEthernet0/0/24]port trunk pvid vlan 100    //G0/0/24 默认 Vlan1
[SW1-GigabitEthernet0/0/24]port trunk allow-pass vlan all
[SW1-GigabitEthernet0/0/24]quit
[SW1]interface Vlanif 10
[SW1-Vlanif10]ip address 172.16.10.1 24
[SW1-Vlanif10]interface Vlanif 20
[SW1-Vlanif20]ip address 172.16.20.1 24
[SW1-Vlanif20]interface Vlanif 100
[SW1-Vlanif100]ip address 172.16.1.1 24
[SW1-Vlanif100]quit
[SW1]
```

② 路由配置。

```
[R1]ospf 1
[R1-ospf-1]area 0
[R1-ospf-1-area-0.0.0.0]network 172.16.1.0 0.0.0.255
[R1-ospf-1-area-0.0.0.0]quit
[R1-ospf-1]quit
[R1]ip route-static 0.0.0.0 0.0.0.0 116.64.64.1
[R1]
------------------------------------------------------------------------------------------
[SW1]ospf 1
[SW1-ospf-1]area 0
[SW1-ospf-1-area-0.0.0.0]network 172.16.1.0 0.0.0.255
[SW1-ospf-1-area-0.0.0.0]network 172.16.10.0 0.0.0.255
[SW1-ospf-1-area-0.0.0.0]network 172.16.20.0 0.0.0.255
[SW1-ospf-1-area-0.0.0.0]quit
[SW1-ospf-1]quit
[SW1]ip route-static 0.0.0.0 0.0.0.0 172.16.1.2
[SW1]
```

(3) Baidu 公司内网配置。

请读者根据网络拓扑自行配置 Baidu 公司内网，让 Web 服务器和 FTP 服务器能够连通

IP 地址 121.32.32.100。

(4) 在路由器 R2～R5 配置 EBGP 路由。

```
[R2]bgp 200                                        //200 为 AS 号,非进程号
[R2-bgp]peer 201.201.201.2 as-number 300  //建立邻居,peer 为对等体
[R2-bgp]peer 204.204.204.2 as-number 400
[R2-bgp]network?                                   //查询 BGP 协议中 network 命令用途
  network Specify a network to announce via BGP   //逐条将 IP 路由表中已经存在的路由条目
                                                            引入 BGP 路由表中
[R2-bgp]network 116.64.64.0 ?
  INTEGER<0-32>      Length of IP address mask
  IP_ADDR<X.X.X.X>  Specify an IPv4 network mask
  route-policy       Specify a route policy
  <cr>               Please press ENTER to execute command
[R2-bgp]network 116.64.64.0 255.255.255.0   //BGP 引入路由后面是网络掩码,即一条完整的路
                                                  由条目,非反网络掩码。其中网络掩码可以省略
[R2-bgp]network 201.201.201.0 24
[R2-bgp]network 204.204.204.0
[R2-bgp]quit
[R2]
```

注意:

① R2 无法与 R5 建立 peer 对等体(邻居)关系,因为 R2 和 R5 之间无法连通,缺少内部路由。即路由器建立对等体关系,前提是能够相互连通。

② 在 IGP 内部网关协议中(如 RIP、OSPF),network 命令用于宣告接口直连网段,后面接反网络掩码表示匹配规则;在 BGP 外部网关协议中,network 命令用于将一条路由条目(该条目必须存在于路由表中,可通过 display ip routing-table 查看)注入 BGP 路由表中。

例如,查看路由表如下(该例与本工作任务无关):

```
[Huawei]display ip routing-table
Route Flags: R - relay, D - download to fib
------------------------------------------------------------------------------
Routing Tables: Public
         Destinations : 14        Routes : 14
  Destination/Mask    Proto  Pre  Cost      Flags NextHop         Interface
     116.64.64.0/24  Direct 0    0           D   116.64.64.100   Serial2/0/0
     116.64.64.1/32  Direct 0    0           D   116.64.64.1     Serial2/0/0
   116.64.64.100/32  Direct 0    0           D   127.0.0.1       Serial2/0/0
   116.64.64.255/32  Direct 0    0           D   127.0.0.1       Serial2/0/0
       127.0.0.0/8   Direct 0    0           D   127.0.0.1       InLoopBack0
       127.0.0.1/32  Direct 0    0           D   127.0.0.1       InLoopBack0
 127.255.255.255/32  Direct 0    0           D   127.0.0.1       InLoopBack0
      172.16.1.0/24  Direct 0    0           D   172.16.1.2      GigabitEthernet0/0/0
      172.16.1.2/32  Direct 0    0           D   127.0.0.1       GigabitEthernet0/0/0
    172.16.1.255/32  Direct 0    0           D   127.0.0.1       GigabitEthernet0/0/0
     172.16.10.0/24  OSPF   10   2           D   172.16.1.1      GigabitEthernet0/0/0
     172.16.20.0/24  OSPF   10   2           D   172.16.1.1      GigabitEthernet0/0/0
 255.255.255.255/32  Direct 0    0           D   127.0.0.1       InLoopBack0

[Huawei]bgp 100
```

```
[Huawei-bgp]network 172.16.10.0 255.255.255.0  //将 OSPF 路由条目<172.16.10.0>引入
                                                 BGP 路由表中以向对等体宣告
```

③ BGP 不产生路由，也不发现路由，只是将 BGP 路由表(非 IP 路由表，可通过脚本 dis bgp routing-table 查看)中路由条目选择性宣告给对等体(可以认为只是转发通告，并不鉴别真伪)。其中，作为 EBGP 对等体，通告本地引入路由、学习到的 BGP 路由(包括 IBGP 和 EBGP 路由)；作为 IBGP 对等体，只通告本地 BGP 引入路由(引入包括 network 或 import-route)和学习到的 EBGP 路由，不通告学习到的 IBGP 路由，详细请参考工作任务十三。

```
---------------------------------------------------------------------------
[R3]bgp 300
[R3-bgp]peer 201.201.201.1 as-number 200
[R3-bgp]peer 202.202.202.1 as-number 500
[R3-bgp]import- route ?
  direct Connected routes
  isis  Intermediate System to Intermediate System (IS- IS) routes
  ospf  Open Shortest Path First (OSPF) routes
  rip  Routing Information Protocol (RIP) routes
  static Static routes
  unr  User network routes
[R3-bgp]import-route direct  //同时引入多条直连路由至 BGP 路由表，避免使用 network 命令
                               逐条引入的烦琐
[R3-bgp]quit
[R3]
---------------------------------------------------------------------------
[R4]bgp 400
[R4-bgp]peer 203.203.203.1 as-number 500
[R4-bgp]peer 204.204.204.1 as-number 200
[R4-bgp]import-route direct
[R4-bgp]quit
[R4]
---------------------------------------------------------------------------
[R5]bgp 500
[R5-bgp]peer 202.202.202.2 as-number 300
[R5-bgp]peer 203.203.203.2 as-number 400
[R5-bgp]import-route direct
[R5-bgp]quit
[R5]
```

(5) 在路由器 R1 和 R6 上配置 Easy-IP 和 nat server。

```
[R1]acl 2000
[R1-acl-basic-2000]rule permit source any
[R1-acl-basic-2000]quit
[R1]interface Serial 2/0/0
[R1-Serial2/0/0]nat outbound 2000
[R1-Serial2/0/0]quit
[R1]
---------------------------------------------------------------------------
[R6]acl 2000
[R6-acl-basic-2000]rule permit source any
[R6-acl-basic-2000]quit
```

```
[R6]interface Serial 2/0/0
[R6-Serial2/0/0]nat outbound 2000
[R6-Serial2/0/0]nat server protocol tcp global current-interface 80 inside 192.168.
10.10 80
[R6-Serial2/0/0]nat server protocol tcp global current-interface 21 inside 192.168.
20.10 21
[R6-Serial2/0/0]quit
[R6]
```

注意：如果只配置 Nat Server，不配置 Easy-IP，仍不能发布内网 Baidu Web 和 FTP 站点，因为服务器不能访问 Internet。

2．任务验证

(1) EBGP 邻居关系验证。

以路由器 R2 为例，R2 分别和路由器 R3(201.201.201.2)、R4(204.204.204.2)成功建立邻居，邻居状态为 Established。

```
[R2]display bgp peer
 BGP local router ID : 201.201.201.1
 Local AS number : 200
 Total number of peers : 2 Peers in established state : 2
  Peer            V  AS    MsgRcvd  MsgSent  OutQ  Up/Down     State        PrefRcv
  201.201.201.2   4  300     68       74        0 00:52:07    Established      5
  204.204.204.2   4  400     49       50        0 00:33:19    Established      5
```

注：请读者自行在路由器 R3、R4 和 R5 上查看所建立的 EBGP 邻居关系，务必为 Established 状态。如发现为 ldle 初始化状态或 Connect 正在连接状态，说明邻居接口之间 TCP 无法建立会话连接，应检查路由器之间接口是否可以 ping 通。

(2) 查看 IP 路由表和 BGP 路由表。

以路由器 R2 为例，查看 IP 路由表如下：

```
[R2]display ip routing-table
Route Flags: R - relay, D - download to fib
------------------------------------------------------------------------------
Routing Tables: Public
         Destinations : 18        Routes : 18
 Destination/Mask    Proto  Pre  Cost     Flags NextHop         Interface
    116.64.64.0/24  Direct 0    0          D   116.64.64.1     Serial2/0/0
    116.64.64.1/32  Direct 0    0          D   127.0.0.1       Serial2/0/0
  116.64.64.100/32  Direct 0    0          D   116.64.64.100   Serial2/0/0
  116.64.64.255/32  Direct 0    0          D   127.0.0.1       Serial2/0/0
    121.32.32.0/24  EBGP   255  0          D   201.201.201.2   GigabitEthernet0/0/0
  121.32.32.100/32  EBGP   255  0          D   201.201.201.2   GigabitEthernet0/0/0
      127.0.0.0/8   Direct 0    0          D   127.0.0.1       InLoopBack0
      127.0.0.1/32  Direct 0    0          D   127.0.0.1       InLoopBack0
127.255.255.255/32  Direct 0    0          D   127.0.0.1       InLoopBack0
  201.201.201.0/24  Direct 0    0          D   201.201.201.1   GigabitEthernet0/0/0
  201.201.201.1/32  Direct 0    0          D   127.0.0.1       GigabitEthernet0/0/0
201.201.201.255/32  Direct 0    0          D   127.0.0.1       GigabitEthernet0/0/0
  202.202.202.0/24  EBGP   255  0          D   201.201.201.2   GigabitEthernet0/0/0
```

```
  203.203.203.0/24  EBGP   255  0        D   204.204.204.2  GigabitEthernet0/0/1
  204.204.204.0/24  Direct 0    0        D   204.204.204.1  GigabitEthernet0/0/1
  204.204.204.1/32  Direct 0    0        D   127.0.0.1      GigabitEthernet0/0/1
204.204.204.255/32  Direct 0    0        D   127.0.0.1      GigabitEthernet0/0/1
255.255.255.255/32  Direct 0    0        D   127.0.0.1      InLoopBack0
```

可以看到,R2 通过 EBGP 学习到所有非直连网段路由条目。

以 R2 为例,查看的 BGP 路由表如下:

```
[R2]display bgp routing-table
 BGP Local router ID is 201.201.201.1
 Status codes: * - valid, > - best, d - damped,
               h - history, i - internal, s - suppressed, S - Stale
               Origin : i - IGP, e - EGP, ? - incomplete
 Total Number of Routes: 13
      Network            NextHop        MED     LocPrf   PrefVal  Path/Ogn
 *>   116.64.64.0/24     0.0.0.0        0                 0       i
 *>   121.32.32.0/24     201.201.201.2                    0       300 500?
 *                       204.204.204.2                    0       400 500?
 *>   121.32.32.100/32   201.201.201.2                    0       300 500?
 *                       204.204.204.2                    0       400 500?
 *>   201.201.201.0      0.0.0.0        0                 0       i
                         201.201.201.2  0                 0       300?
 *>   202.202.202.0      201.201.201.2  0                 0       300?
 *                       204.204.204.2                    0       400 500?
 *>   203.203.203.0      204.204.204.2  0                 0       400?
 *                       201.201.201.2                    0       300 500?
 *>   204.204.204.0      0.0.0.0        0                 0       i
                         204.204.204.2  0                 0       400?
```

在 BGP 路由表中,黑体字<116.64.64.0>、<201.201.201.0>和<204.204.204.0>网段为通过 network 命令手动引入,i 表示 IBGP;剩余路由条目通过 EBGP 对等体通告后习得。"*>"表示有效路由,"*"表示存在多条路径的次优路由。"300 500"表示该 BGD 路由从 500 区域路由器通告至 300 区域路由器。

(3) 连通性测试。

① 主机 1、主机 2、Baidu Web 和 Baidu FTP 服务器可以连通公网所有路由器。其中主机 1 与 IP 地址 121.32.32.100 连通情况如图 11-2 所示。

② Baidu 公司 Web 服务器访问测试。

在学习站点下载"baidu 网站"文件,在 Server1 发布公司 Web 站点,公司主机 1 通过地址 http://121.32.32.100/index.htm 可以访问 Baidu 公司 Web 服务器站点,如图 11-3 所示。

③ Baidu 公司 FTP 服务器访问测试。

在 Server2 上发布 Baidu 公司 FTP 站点,在 Server2 窗口依次选择"服务器信息"→Ftp Server 自定义站点根目录,单击"启动"按钮,如图 11-4 所示。

注意:FTP 在穿越 NAT 时,应采用被动 PASV 模式,使用传统浏览器(如 IE、chrome 等基于 port 模式)无法访问,需安装第三方 FTP 应用软件(如 FlashFXP、CuteFTP 等)。华为 eNSP 模拟器中 Client 设备无法很好地模拟计算机操作系统,更无法安装应用软件,此时可以

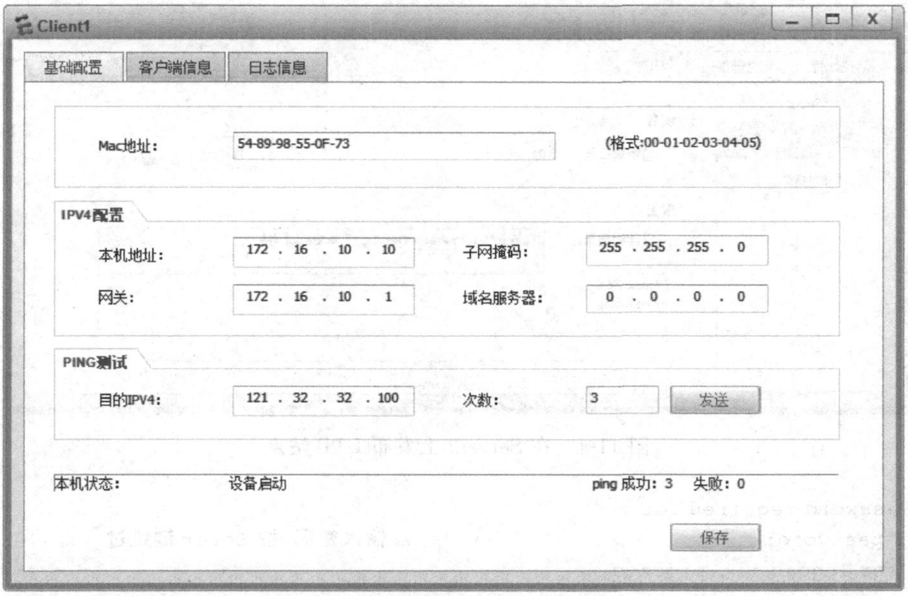

图 11-2 主机 1 与 IP 地址 121.32.32.100 连通

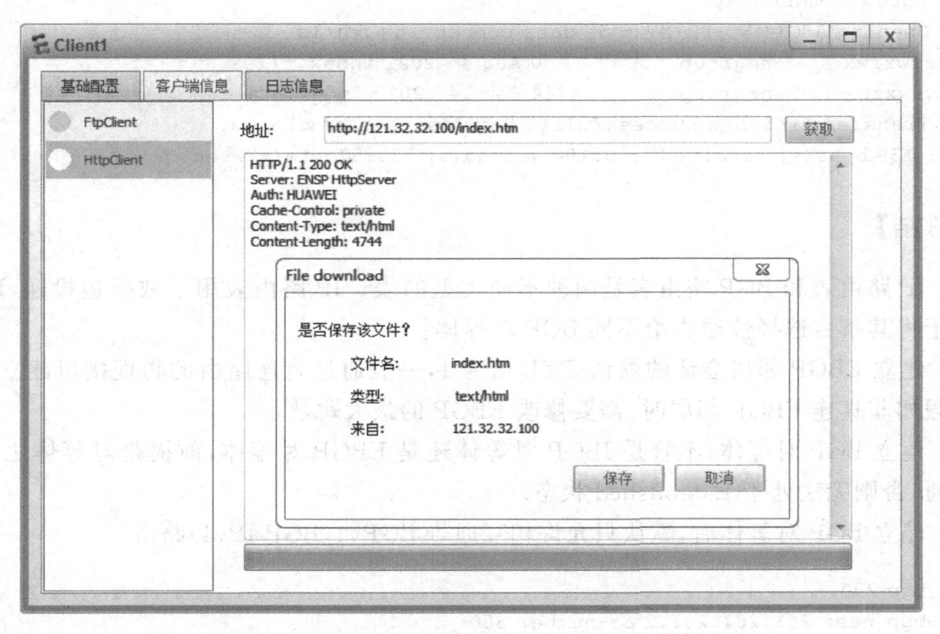

图 11-3 主机 2 可以访问 Server1 Web 站点

通过在路由器 R1 上接入 Baidu 公司 FTP 服务器进行测试。

```
< R1> ftp 121.32.32.100
Trying 121.32.32.100 ...
Press CTRL+K to abort
Connected to 121.32.32.100.
220 FtpServerTry FtpD for free
User(121.32.32.100:(none)):                    //输入账号,按 Enter 键跳过
```

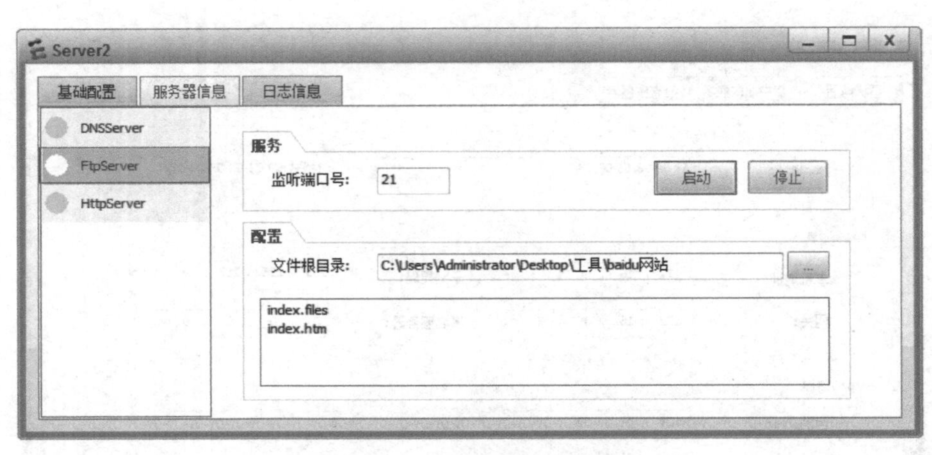

图 11-4 在 Server2 上发布 FTP 站点

```
331 Password required for .
Enter password:                                    //输入密码,按 Enter 键跳过
230 User logged in , proceed

[R1-ftp]dir                                        //dir 命令用于显示当前目录
200 Port command okay.
150 Opening ASCII NO-PRINT mode data connection for ls -l.
drwxrwxrwx 1     nogroup              0 Aug 4   2021 index.files
-rwxrwxrwx 1     nogroup           4744 Feb 14   2012 index.htm
226 Transfer finished successfully. Data connection closed.
FTP: 139 byte(s) received in 0.190 second(s) 731.57byte(s)/sec.
```

【任务总结】

(1) IP 路由表和 BGP 路由表是两种不同类型的表。IP 路由表用于数据包投递,BGP 路由表用于将其条目选择性通告给不同 BGP 对等体。

(2) 建立 EBGP 邻居会话的默认 TTL 值为 1,一般通过直连路由的物理接口建立 EBGP 邻居。假如非直连 EBGP 邻居时,需要修改 EBGP 的最大跳数。

(3) 建立 BGP 对等体,不管是 IBGP 对等体还是 EBGP 对等体,前提是对等体之间能够相互连通,否则无法处于 Established 状态。

(4) 建立 BGP 对等体后,默认只允许相互通告 BGPv4(BGP IPv4)路由。

```
[R2]bgp 200
[R2-bgp]peer 201.201.201.2 as-number 300
[R2-bgp]peer 201.201.201.2 enable   //让 201.201.201.2 成为其对等体,并相互通告 BGPv4 路
                                      由。本行可以不输入
```

工作任务十二

BGP 路径选择

【工作目的】

掌握修改 BGP 属性 Preferred Value 值实现流量分担。

【工作任务】

运营商路由器通过 BGP 协议互联，组成公网拓扑，为公司 A、Baidu 公司、Sina 公司提供网络接入服务。运营商管理员测试发现，路由器 R2 与 R5 之间流量全部经由 R3 转发，R4 处于闲置状态。为合理利用线路带宽，须采用路由策略将去往 Baidu 公司的流量引至 R4，达到负载均衡的目的，具体任务如下。

(1) 公司 A 去往 Sina 公司流量经 R3 转发。

(2) 公司 A 去往 Baidu 公司流量经 R4 转发。

(3) 配置 Nat Server，主机 1 和主机 2 可以访问 Sina 公司和 Baidu 公司的 Web 服务与 FTP 服务。

【工作背景】

当一台 BGP 路由器中存在多条去往同一目标网络的 BGP 路由时，BGP 会对这些 BGP 路由属性进行比较，以确定去往该目标网络的最优 BGP 路由，然后将该最优 BGP 路由与去往同一目标网络的其他协议路由进行比较，从而决定是否将该最优 BGP 路由放进 IP 路由表中。在对 BGP 路由属性进行比较时，BGP 会遵循一定的先后次序，直到确定出一条最优路由为止。在 BGP 路由属性的比较过程中，首先要比较的就是路由信息选值 Preferred Value，也简称为 PrefVal。

路由信息的首选值 Preferred Value 的取值范围是 0～65535，取值越大，优先级越高。默认情况下，Preferred Value 取值为 0；通过修改 Preferred Value 的值，可以很方便地实现对路径选择的控制。

【任务分析】

1. 路径择优顺序

当一台 BGP 路由器中存在多条去往同一目标网络的 BGP 路由时，按以下顺序选择最优路径。

(1) preferred-value 值：路由信息的首选值，值越大越优，默认为 0，是华为的私有属性，具有本地意义，不会随路由条目传递。

(2) local-preference 值：值越大越优，其中 IBGP 通告时值默认为 100(EBGP 不通告

local-preference 值)。

(3) 路由生成方式：network>import-route。

(4) AS_Path 属性：路径总长度(AS 相加和)最短的路径优先。

(5) Origin 属性：比较 Origin 属性,IGP 优于 EGP,EGP 优于 incomplete。

(6) MED 属性：选择 MED 值较小的路由,默认只比较来自同一 AS 的 MED 值。

(7) BGP 对等体类型：即比较 IBGP 邻居或 EBGP 邻居,EBGP 路由优先于 IBGP 路由(通过内部优先级比较,EBGP 为 20,IBGP 为 200)。

(8) BGP 优先选择到 BGP 下一跳的 IGP 的 cost 值最小的一个。

当以上全部相同,则为等价路由,再按照以下顺序比较。

(9) 比较 Cluster-list 长度,短者优先(RR)。

(10) 比较 Originator_ID(如果没有 Originator_ID,则用 Router ID 比较),选择数值较小的路径。

(11) 比较对等体 IP 地址(Router ID 值),选择 IP 地址数值最小的路径。

2. 地址前缀列表与过滤规则

地址前缀列表是一种包含一组路由信息过滤规则的过滤方式。地址前缀列表可以匹配路由的网络号及掩码,当待过滤的路由已匹配当前表项的网络号时,掩码长度可以用于进行精确匹配或者在一定掩码长度范围内匹配。IP 地址前缀参数见表 12-1。

地址前缀列表过滤路由的原则可以总结为：顺序匹配、唯一匹配、默认拒绝。因此在一个地址前缀列表中创建了一个或多个 deny 模式的表项后,需要创建一个表项来允许所有其他路由通过。如拒绝(10.0.0.0)网段访问公网,允许其他网段访问公网,地址前缀列表定义如下：

```
[Huawei]ip ip-prefix deny_to_internet 10.0.0.0 8 greater-equal 8
[Huawei]ip ip-prefix deny_to_internet index 20 permit 0.0.0.0 0 less-equal 32
```

其语法规则如下,各主要参数见表 12-1。

```
ip ip-prefix ip-prefix-name [index index-number] {permit|deny} ipv4-address mask-
length [greater-equal greater-equal-value] [less-equal less-equal-value]
```

表 12-1 IP 地址前缀参数表

参　　数	含　　义
ipv4-address	用于指定网络号
mask-length	用于限定网络号的前多少位需严格匹配
greater-equal greater-equal-value	表示掩码大于或等于 greater-equal-value
less-equal less-equal-value	表示掩码小于或等于 less-equal-value

注：

(1) 如果不配置 greater-equal 和 less-equal 参数,则进行精确匹配,即只匹配掩码长度为 mask-length 的路由。

(2) 如果只配置参数 greater-equal,则匹配的掩码长度范围为[greater-equal-value,32]。

(3) 如果只配置参数 less-equal,则匹配的掩码长度范围为[mask-length,less-equal-value]。

(4) 如果同时配置参数 greater-equal 和 less-equal,则匹配的掩码长度范围为[greater-equal-value,less-equal-value]。

3. 路由策略 route-policy

路由策略是为改变网络流量所经过的途径而对路由信息采用的方法，该策略能够影响到路由产生、发布、选择等，进而影响数据包的转发路径。工具包括 ACL、ip-prefix、filter-policy 等，方法包括对路由进行过滤、设置路由的属性等。路由策略的作用见表 12-2。

表 12-2 路由策略作用表

作　用	执 行 过 程	作　用
对路由信息进行过滤	如果某条路由符合某种条件，那么则接收这条路由。 如果某条路由符合某种条件，那么则发布这条路由。 如果某条路由符合某种条件，那么则引入这条路由	控制路由的接收、发布和引入，提高网络安全性
修改路由项属性值	如果某条路由项符合某种条件，那么将这条路由的某个属性值修改为一个特定值	修改路由属性，对网络数据流量进行合理规划，提高网络性能

以下为创建路由策略的语法。

```
route-policy route-policy-name { permit | deny } node nodeid
```

当使用 Route-Policy 时，node 的值小的节点先进行匹配。一个节点匹配成功后，路由将不再匹配其他节点。

Route-Policy 由节点号、匹配模式、if-match 子句（条件语句）和 apply 子句（执行语句）四个部分组成。

(1) 节点号。

一个 Route-Policy 可以由多个节点（node）构成。路由匹配 Route-Policy 时遵循以下两个规则。

① 顺序匹配：在匹配过程中，系统按节点号从小到大的顺序依次检查各个表项，因此在指定节点号时，要注意符合期望的匹配顺序。

② 唯一匹配：Route-Policy 各节点号之间是“或”的关系，只要通过一个节点的匹配，就认为通过该过滤器，不再进行其他节点的匹配。

(2) 节点的匹配模式有两种：permit 和 deny。

参数 permit 指定节点的匹配模式为允许，即允许满足匹配条件的路由信息（路由条目）通过并更改参数属性。当路由项通过该节点的过滤后，将执行该节点的 apply 子句，不进入下一个节点；如果路由项没有通过该节点过滤，将进入下一个节点继续匹配。

参数 deny 指定节点的匹配模式为拒绝，即不允许满足匹配条件的路由信息（路由条目）通过。当路由项满足该节点的所有 if-match 子句时，将被拒绝通过该节点，不进入下一个节点；如果路由项不满足该节点的 if-match 子句，将进入下一个节点继续匹配。

(3) if-match 子句（条件语句）。

if-match 子句用来定义一些匹配条件。Route-Policy 语句的每一个节点可以含有多个 if-match 子句，也可以不含 if-match 子句。如果某个 permit 节点没有配置任何 if-match 子句，则该节点匹配所有路由。

(4) apply 子句（执行语句）。

apply 子句用来指定动作。路由通过 Route-Policy 语句过滤时，系统按照 apply 子句指定的动作对路由信息的一些属性进行设置。Route-Policy 语句的每一个节点可以含有多个

apply 子句，也可以不含 apply 子句。如果只需要过滤路由，不需要设置路由的属性，则不使用 apply 子句。

【设备器材】

三层交换机(S5700)3 台；路由器(AR1220)7 台，部分须添加 1GEC 千兆接口模块或 2SA 串口模块。

主机 6 台，承担角色见表 12-3。

表 12-3　主机配置表

角　　色	接 入 方 式	IP 地址	所 属 公 司
主机 1	eNSP PC 接入	172.16.10.10/24	公司 A 技术部
主机 2	eNSP PC 接入	172.168.20.10/24	公司 A 工程部
Baidu Web 服务器	eNSP PC 接入	192.168.10.10/24	Baidu 公司
Baidu FTP 服务器	eNSP PC 接入	192.168.20.10/24	Baidu 公司
Sina Web 服务器	eNSP PC 接入	192.168.10.10/24	Sina 公司
Sina FTP 服务器	eNSP PC 接入	192.168.20.10/24	Sina 公司

【环境拓扑】

环境拓扑如图 12-1 所示。

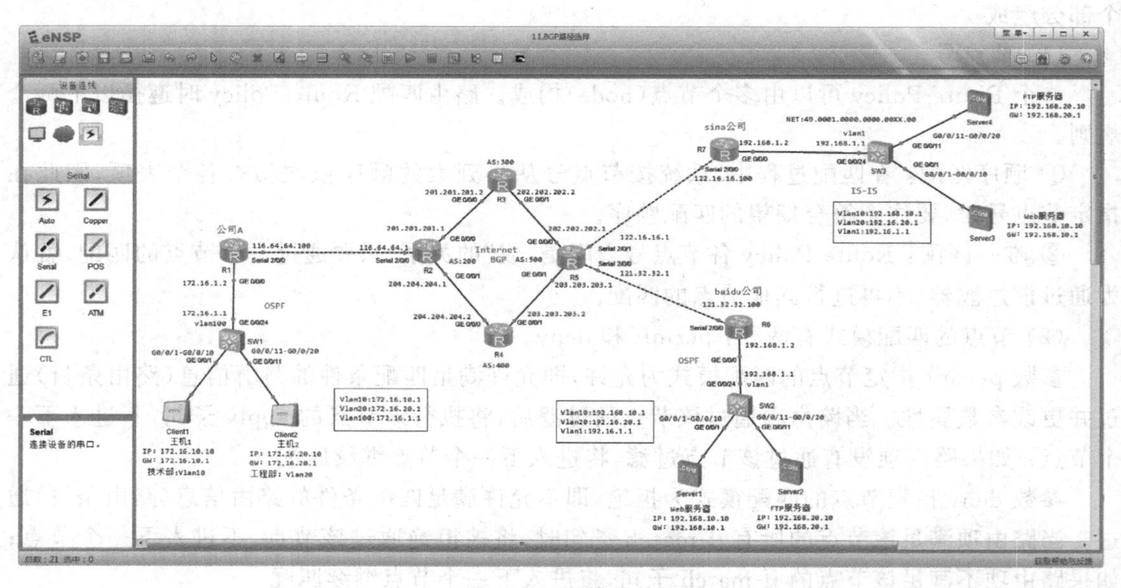

图 12-1　环境拓扑

【工作过程】

1. 基本配置

(1) 路由器接口 IP 和系统名称配置。

请读者根据网络拓扑自行配置路由器接口 IP 地址和系统名称，注意所有接口子网掩码长度均为 24。

(2) A 公司、Baidu 公司内网配置。

请读者根据网络拓扑自行配置,也可以将工作任务十一中的相关配置导入设备。

(3) Sina 公司内网配置。

① Vlan 划分与 IP 地址配置。

```
[SW3]vlan batch 10 20
[SW3]port-group 1
[SW3-port-group-1]group-member GigabitEthernet 0/0/1 to GigabitEthernet 0/0/10
[SW3-port-group-1]port link-type access
[SW3-port-group-1]port default vlan 10
[SW3-port-group-1]quit
[SW3]port-group 2
[SW3-port-group-2]group-member GigabitEthernet 0/0/11 to GigabitEthernet 0/0/20
[SW3-port-group-2]port link-type access
[SW3-port-group-2]port default vlan 20
[SW3-port-group-2]quit
[SW3]interface GigabitEthernet 0/0/24
[SW3-GigabitEthernet0/0/24]port link-type trunk
[SW3-GigabitEthernet0/0/24]port trunk allow-pass vlan all
[SW3-GigabitEthernet0/0/24]quit
[SW3]interface Vlanif 10
[SW3-Vlanif10]ip address 192.168.10.1 24
[SW3-Vlanif10]interface Vlanif 20
[SW3-Vlanif20]ip address 192.168.20.1 24
[SW3-Vlanif20]interface Vlanif 1
[SW3-Vlanif1]ip address 192.168.1.1 24
[SW3-Vlanif1]quit
[SW3]
```

② 路由配置。

```
[R7]isis 1
[R7-isis-1]network-entity 49.0001.0000.0000.0001.00
[R7-isis-1]is-level level-1
[R7-isis-1]quit
[R7]interface GigabitEthernet 0/0/0
[R7-GigabitEthernet0/0/0]isis enable 1
[R7-GigabitEthernet0/0/0]quit
[R7]ip route-static 0.0.0.0 0.0.0.0 122.16.16.1
[R7]
--------------------------------------------------------------------------------
[SW3]isis 1
[SW3-isis-1]network-entity 49.0001.0000.0000.0010.00
[SW3-isis-1]is-level level-1
[SW3]interface Vlanif 1
[SW3-Vlanif1]isis enable 1
[SW3-Vlanif1]interface Vlanif 10
[SW3-Vlanif10]isis enable 1
[SW3-Vlanif10]interface Vlanif 20
[SW3-Vlanif20]isis enable 1
```

```
[SW3-Vlanif20]quit
[SW3]ip route-static 0.0.0.0 0.0.0.0 192.168.1.2
[SW3]
```

(4) 在路由器 R2～R5 上配置 EBGP 路由。

请读者根据网络拓扑自行配置,也可以将工作任务十一中的相关配置导入设备。

(5) 在路由器 R1、R6 和 R7 上配置 Easy-IP 和 Nat Server。

请读者根据网络拓扑自行配置,也可以将工作任务十一中的相关配置导入设备。

(6) 任务测试。

在路由器 R2 上查看 BGP 路由表。

```
[R2]display bgp routing-table
 BGP Local router ID is 201.201.201.1
 Status codes: * - valid, > - best, d - damped,
               h - history, i - internal, s - suppressed, S - Stale
               Origin : i - IGP, e - EGP, ? - incomplete
 Total Number of Routes: 17
      Network            NextHop         MED     LocPrf   PrefVal  Path/Ogn
 *>   116.64.64.0/24     0.0.0.0         0                0        i
 *>   121.32.32.0/24     201.201.201.2                    0        300 500?
 *                       204.204.204.2                    0        400 500?
 *>   121.32.32.100/32   201.201.201.2                    0        300 500?
 *                       204.204.204.2                    0        400 500?
 *>   122.16.16.0/24     201.201.201.2                    0        300 500?
 *                       204.204.204.2                    0        400 500?
 *>   122.16.16.100/32   201.201.201.2                    0        300 500?
 *                       204.204.204.2                    0        400 500?
 *>   201.201.201.0      0.0.0.0         0                0        i
                         201.201.201.2   0                0        300?
 *>   202.202.202.0      201.201.201.2   0                0        300?
 *                       204.204.204.2                    0        400 500?
 *>   203.203.203.0      204.204.204.2   0                0        400?
 *                       201.201.201.2                    0        300 500?
 *>   204.204.204.0      0.0.0.0         0                0        i
                         204.204.204.2   0                0        400?
```

从以上路由表可以看到以下情况。

① 由于 R5 引入直连路由至 BGP 路由表,通过 EBGP 通告,R2 的 BGP 表中出现＜121.32.32.0/24＞、＜121.32.32.100/32＞、＜122.16.16.0/24＞和＜122.16.16.100/32＞四个网段。查看 R5 的 IP 路由表如下:

```
[R5]display ip routing-table
Route Flags: R - relay, D - download to fib
------------------------------------------------------------------------------
Routing Tables: Public
         Destinations : 21       Routes : 21
  Destination/Mask  Proto  Pre  Cost      Flags NextHop         Interface

     116.64.64.0/24  EBGP   255  0           D  202.202.202.2   GigabitEthernet0/0/0
     121.32.32.0/24  Direct 0    0           D  121.32.32.1     Serial2/0/0
   121.32.32.100/32  Direct 0    0           D  121.32.32.100   Serial2/0/0
```

```
   122.16.16.0/24  Direct 0    0        D    122.16.16.1       Serial2/0/1
  122.16.16.100/32 Direct 0    0        D    122.16.16.100     Serial2/0/1
  ...
  201.201.201.0/24 EBGP    255  0        D    202.202.202.2     GigabitEthernet0/0/0
  204.204.204.0/24 EBGP    255  0        D    203.203.203.2     GigabitEthernet0/0/1
255.255.255.255/32 Direct 0    0        D    127.0.0.1         InLoopBack0
```

② 在 R2 的 BGP 路由表中，去往(121.32.32.0/24)、(121.32.32.100/32)、(122.16.16.0/24)和(122.16.16.100/32)四个网段存在两条路径，经 R3 转发的下一跳 201.201.201.2 或经 R4 转发的下一跳 204.204.204.2，但是最终 R2 优选的都是下一跳为 201.201.201.2 路由。

当一个路由器收到两条相同目的地的路由时，会依次比较路由信息首选值 PreVal、本地优先级 LocPrf、路由生成方式、AS_Path 属性、Origin 属性、MED 属性、BGP 对等体类型等参数。

其中路由信息首选值 PreVal 均默认为 0，本地优先级 LocPrf 默认值均为 100，路由生成方式均为直连路由引入，接着比较 AS_Path 属性，BGP 会以 AS 路径最短的路由作为最优路由。其中，经(201.201.201.2)路由器 R3 转发的 AS_Path 为 300＋500，经＜204.204.204.2＞路由器 R4 转发的 AS_Path 为 400＋500，因此最终 R2 选择途经 R3 发布的路由作为最优路由。

③ R4 没有分担任何流量，次优路由仅作为备份路径，造成链路带宽浪费，只有当 R3 发生故障时才能承接业务流量。此时可以在 R2 上修改 PreVal(preferred value)值，使得 R4 分担从 R2 去往(121.32.32.0/24)、(121.32.32.100/32)的流量。

(7) 修改 R2 的 Preferred Value 值。

```
[R2]ip ip-prefix to_baidu index 10 permit 121.32.32.0 24   //创建 IPv4 地址前缀列表
```

注意：前缀列表通过前缀(网段)和前缀长度(网络掩码)匹配一个网段。

① to_baidu：前缀列表名，可以是名字或者数字，取值 STRING＜1～169＞。

② index 参数：索引号(序列号)，取值 INTEGER＜1～4294967295＞。默认情况下，该序号值按照配置先后顺序依次递增，每次增加 10，第一个序号值为 10。

③ 121.32.32.0：前缀网段。

④ 24：前缀长度，即网络掩码长度。

```
[R2]ip ip-prefix to_baidu index 20 permit 121.32.32.100 32  //根据 R2 路由表和路由转发
                                                              最长匹配原则，必须输入
                                                              此行
[R2]route-policy R2_to_R4 permit node 10   //创建路由策略节点 10
```

注意：路由策略是一种依据用户制订的策略进行路由选择的机制，其优先级要高于普通路由，用于路由器(从对等体)学习路由条目或(向对等体)发布路由条目时如何修改路由条目属性。策略表是若干节点(node)的集合。Route-Policy 由节点号、匹配模式、if-match 子句(条件语句)和 apply 子句(执行语句)这四个部分组成。

① R2_to_R4。路由策略名，可以是名字或者数字，取值 STRING＜1～40＞。

② permit/deny。节点匹配模式有两种：permit 和 deny。

- permit 指定节点的匹配模式为允许。当路由项通过该节点的过滤后，将执行该节点的 apply 子句，不进入下一个节点，如果路由项没有通过该节点过滤，将进入下一个节点继续匹配。

根据上下文,Permit 此时是指允许 R2 允许接收来自对等体 R4 通告的 EBGP<121.32.32.0>和<121.32.32.100>路由项,并更改 preferred-value 参数值。

- deny 指定节点的匹配模式为拒绝,即不允许满足匹配条件的路由项通过。如为 deny,根据上下文,意思是 R2 禁止接收来自对等体 R4 通告的 EBGP<121.32.32.0>和<121.32.32.100>路由项。

③ node。每个策略节点由 node-number 来指定,node-number 值越小优先级越高,优先级高的策略优先被执行。node 取值 INTEGER<0～65535>。

```
[R2-route-policy]if-match ip-prefix to_baidu     //if-match 定义该节点匹配规则。如果满足匹配条件,则允许路由项通过,并执行相应动作 apply 子句。如不满足 if-match 子句,则会进入下一节点 node 20 进行匹配。
[R2-route-policy]apply preferred-value 100       //apply 定义路由项通过该节点过滤后进行的动作
```

注:apply 相应动作有以下常用参数。

```
• apply ip-address next-hop IP_ADDR<X.X.X.X>    //指定下一跳 IP
• apply cost INTEGER<0-4294967295>              //指定路由开销
• apply preferred-value INTEGER<0-65535>        //指定 BGP 路由信息首选值 preferred-value,取值 INTEGER<0~65535>,默认为 0

[R2-route-policy]quit
[R2]route-policy R2_to_R4 permit node 20        //定义路由策略 R2_to_R4 过滤节点 20。节点 20 没有 if-match 语句进行匹配,则表示所有路由项都满足匹配条件,即 permit 允许路由项通过。本行的目的是放行其他路由条目,如放行(122.16.16.0)等路由项
```

注:

① 如果路由项不能通过一个策略所有节点的匹配,则默认拒绝。即一个路由条目,既不满足 permit 放行条件,也不满足 deny 拒绝通过条件,则默认 deny 拒绝通过,类似访问控制列表 ACL。

② 默认所有未匹配的路由将被拒绝通过 Route-Policy,无法通告至其对等体。通常应在多个过滤节点后设置一个不含 if-match 子句的 permit 模式 Route-Policy,用于允许其他所有路由条目通过。

```
[R2-route-policy]quit
[R2]bgp 200
[R2-bgp]peer 204.204.204.2 route-policy R2_to_R4 import
//对来自对等体 R4(204.204.204.2)发布的 BGP 路由加载路由策略。R4 向 R2 发布能去往<121.32.32.0 24>和<121.32.32.100 32>的 BGP 路由条目,R2 则在该条目的 preferred-value 属性值+100,从而影响 R2 择优策略。注意,切勿同时在语句 peer 201.201.201.2 加载路由策略 route-policy R2_to_R4 import,否则会造成通过 R3 去往(121.32.32.0 24)和(121.32.32.100 32)网段的 preferred-value 值同样为 100,最终通过 AS_Path 值决定所有路由条目都经 R3 转发。
  ① import:接收对等体满足条件的路由;
  ② export:向对等体发布满足条件的路由
[R2-bgp]quit
[R2]
```

2. 任务验证

(1) 在路由器R2上重新查看BGP路由表。

```
[R2]dis bgp routing-table
 BGP Local router ID is 201.201.201.1
 Status codes: *  -  valid, >  -  best, d - damped,
               h - history, i -  internal, s - suppressed, S - Stale
               Origin : i -  IGP, e -  EGP, ? -  incomplete
 Total Number of Routes: 17
      Network            NextHop        MED      LocPrf  PrefVal   Path/Ogn
 *>   116.64.64.0/24     0.0.0.0        0                0          i
 * >   121.32.32.0/24    204.204.204.2                   100        400 500?
 *                       201.201.201.2                   0          300 500?
 * >   121.32.32.100/32  204.204.204.2                   100        400 500?
 *                       201.201.201.2                   0          300 500?
 *>   122.16.16.0/24     201.201.201.2                   0          300 500?
 *                       204.204.204.2                   0          400 500?
 *>   122.16.16.100/32   201.201.201.2                   0          300 500?
 *                       204.204.204.2                   0          400 500?
 *>   201.201.201.0      0.0.0.0        0                0          i
                         201.201.201.2  0                0          300?
 *>   202.202.202.0      201.201.201.2  0                0          300?
 *                       204.204.204.2                   0          400 500?
 *>   203.203.203.0      204.204.204.2  0                0          400?
 *                       201.201.201.2                   0          300 500?
 *>   204.204.204.0      0.0.0.0        0                0          i
                         204.204.204.2  0                0          400?
```

从R2的BGP路由表可以看到,R2去往(121.32.32.0 24)和(121.32.32.100 32)网段时,下一跳为路由器R4的IP地址204.204.204.2,从而达到流量分担的目的。

(2) 在R1上验证数据包转发路径。

```
[R1]tracert 122.16.16.100
traceroute to 122.16.16.100(122.16.16.100), max hops: 30 ,packet length: 40,pr
ess CTRL_C to break
 1 116.64.64.1 40 ms **
 2 201.201.201.2 30 ms **                        //经 R3 转发
 3 202.202.202.1 30 ms **
 4 122.16.16.100 40 ms **

[R1]tracert 121.32.32.100
traceroute to 121.32.32.100(121.32.32.100), max hops: 30 ,packet length: 40,pr
ess CTRL_C to break
 1 116.64.64.1 30 ms **
 2 204.204.204.2 30 ms **                        //经 R4 转发
 3 203.203.203.1 50 ms **
 4 121.32.32.100 50 ms **
```

(3) 应用服务器访问测试。

主机1和主机2可以访问Sina公司和Baidu公司的Web服务与FTP服务,请读者自行验证。

【任务总结】

(1) preferred-value 默认值为 0。

(2) 策略路由 PBR(policy-based routing)与路由策略(routing policy)存在以下不同。

① 策略路由的操作对象是数据包,在路由表已经产生的情况下,不按照路由表进行转发,而是根据需要,依照某种策略改变数据包转发的下一条。

② 路由策略的操作对象是路由信息(路由条目)。利用路由策略主要实现了路由过滤和路由属性设置等功能,可通过改变路由属性(包括可达性)来改变网络流量所经过的路径。

(3) 默认所有未匹配的路由条目将被拒绝通过 Route-Policy。如果 Route-Policy 中定义了一个以上的节点,为避免路由条目被节点过滤,通常在多个 deny 节点后设置一个不含 if-match 子句的 permit 模式 Route-Policy,用于允许其他所有路由条目通过。

工作任务十三

IBGP 与 EBGP 通告、引入与过滤

【工作目的】

掌握利用 IBGP 与 EBGP 建立邻居的方式，理解 BGP 路由条目宣告、引入与过滤。

【工作任务】

移动和电信运营商组成公网拓扑。某新就职网络运维人员对 BGP 协议不熟悉，以为动态路由协议可以自动发现路由，对运营商内网路由器引入直连网段后导致部分网络无法连通。为解决这个问题，在不改变物理拓扑前提下对网络重新规划，具体任务如下。

(1) 理解现有运营商路由器 BGP 配置问题和故障原因。

(2) 对网络重新规划，其中移动内网采用 IS-IS 协议，电信内网采用 OSPF 协议，实现内网底层节点的连通。

(3) 移动和电信边界路由器(R5 与 R6)建立 EBGP 对等体，各自引入 IGP 路由并向对方宣告内部网段。

(4) 将移动内部路由器与其边界路由器 R5 建立 IBGP 对等体，电信内部路由器与其边界路由器 R6 建立 IBGP 对等体，实现不同运营商全网互通。

【工作背景】

运营商属于公网，运行 BGP，建立 BGP 邻居关系的前提是接口必须能够建立 TCP 会话(即能 ping 通)，底层一般通过 OSPF 或 IS-IS 路由实现路由器接口之间连通性。在建立 BGP 对等体时，在相同 AS 内，对于 IBGP 建议采用 LooBack 虚拟接口建立邻居关系，避免物理接口处于 Down 状态导致邻居失效。在不同 AS 内，对于 EBGP 建议采用物理接口建立邻居(EBGP 邻居之间在发送 BGP 报文时，TTL 默认值为 1，所以 EBGP 默认要求邻居之间必须物理直连)。

【任务分析】

1. IBGP 与 EBGP

BGP 邻居关系分为 IBGP(internal BGP)和 EBGP(external BGP)两种。

(1) IBGP。运行在相同 AS 内的 BGP 路由器建立的邻居关系。

(2) EBGP。运行在不同 AS 之间的 BGP 路由器建立的邻居关系。

注意：配置 BGP 对等体(邻居)时使用的 IP 地址，应互为 BGP 报文的源 IP 地址和目的 IP 地址。对于 IBGP 建议采用 LooBack 虚拟接口建立对等体，而对于 EBGP 建议采用直连物理接口建立对等体。

2. BGP 路由引入

BGP 并不产生路由，也不发现路由，因此需要将其他路由(如 IGP 路由等)引入 BGP 路由表中，从而将这些路由在 AS 之内和 AS 之间传播。BGP 支持通过以下两种方式引入路由。

(1) Network。在 IP 路由表中将已经存在的路由项逐条引入 BGP 路由表，比 Import 方式更精确。

(2) Import。按协议类型，将 RIP、OSPF、IS-IS 等路由协议引入 BGP 路由表中。为了保证引入的 IGP 路由的有效性，使用 Import 方式还可以引入静态路由和直连路由。

【设备器材】

三层交换机(S5700)3 台，路由器(AR1220)8 台，部分须添加 1GEC 千兆接口模块或 2SA 串口模块。

【环境拓扑】

环境拓扑如图 13-1 所示。

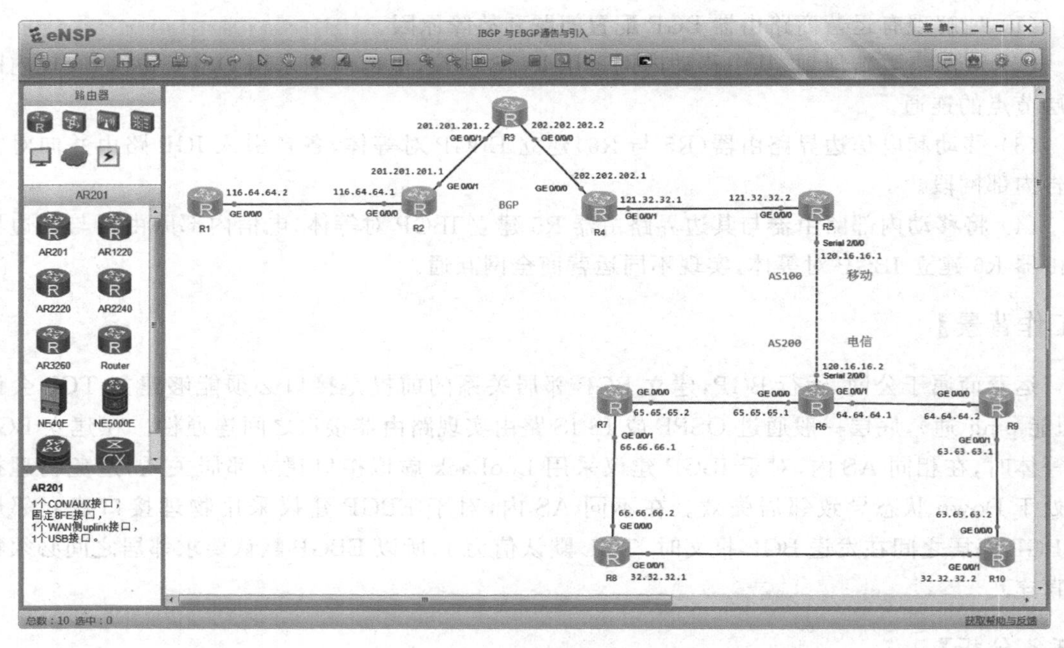

图 13-1 环境拓扑

【工作过程】

1. 基本配置

注：以下是错误配置。

(1) 路由器接口 IP 地址和系统名称配置。

请读者根据网络拓扑自行配置路由器接口 IP 地址和系统名称，注意所有接口子网掩码长度均为 24。

(2) 移动区域 BGP 配置。

```
[R1]bgp 100
[R1-bgp]peer 116.64.64.1 as-number 100        //同一区域属于 IBGP
[R1-bgp]network 116.64.64.0 24
[R1-bgp]quit
[R1]
------------------------------------------------------------------------
[R2]bgp 100
[R2-bgp]peer 116.64.64.2 as-number 100
[R2-bgp]peer 201.201.201.2 as-number 100
[R2-bgp]network 116.64.64.0 24
[R2-bgp]network 201.201.201.0 24
[R2-bgp]quit
[R2]
------------------------------------------------------------------------
[R3]bgp 100
[R3-bgp]peer 201.201.201.1 as-number 100
[R3-bgp]peer 202.202.202.1 as-number 100
[R3-bgp]network 201.201.201.0 24
[R3-bgp]network 202.202.202.0 24
[R3-bgp]quit
[R3]
------------------------------------------------------------------------
[R4]bgp 100
[R4-bgp]peer 202.202.202.2 as-number 100
[R4-bgp]peer 121.32.32.2 as-number 100
[R4-bgp]network 202.202.202.0 24
[R4-bgp]network 121.32.32.0 24
[R4-bgp]quit
[R4]
------------------------------------------------------------------------
[R5]ip route-static 0.0.0.0 0.0.0.0 120.16.16.2
[R5]bgp 100
[R5-bgp]peer 121.32.32.1 as-number 100        //同一区域属于 IBGP
[R5-bgp]peer 120.16.16.2 as-number 200        //不同区域属于 EBGP
[R5-bgp]network 121.32.32.0 24
[R5-bgp]network 120.16.16.0 24                //注意,必须先配置 R6 的 S2/0/0 接口 IP,否则
                                                本地接口 S2/0/0 处于 Down 状态,导致引入网
                                                段出错
[R5-bgp]import-route static                   //引入静态路由。虽然默认路由也属于静态路
                                                由,但 BGP 默认不允许引入静态默认路由
[R5-bgp]default-route imported                //允许 BGP 引入本地 IP 路由表中已经存在的默
                                                认路由
[R5-bgp]quit
[R5]
```

(3) 电信区域 BGP 配置。

```
[R6]ip route-static 0.0.0.0 0.0.0.0 120.16.16.1
[R6]bgp 200
[R6-bgp]peer 120.16.16.1 as-number 100
[R6-bgp]peer 65.65.65.2 as-number 200
```

```
[R6-bgp]peer 64.64.64.2 as-number 200
[R6-bgp]network 120.16.16.0 24
[R6-bgp]network 65.65.65.0 24
[R6-bgp]network 64.64.64.0 24
[R6-bgp]import-route static
[R6-bgp]default-route imported
[R6-bgp]quit
[R6]
--------------------------------------------------------------------------------
[R7]bgp 200
[R7-bgp]peer 65.65.65.1 as-number 200
[R7-bgp]peer 66.66.66.2 as-number 200
[R7-bgp]network 65.65.65.0 24
[R7-bgp]network 66.66.66.0 24
[R7-bgp]quit
[R7]
--------------------------------------------------------------------------------
[R8]bgp 200
[R8-bgp]peer 66.66.66.1 as-number 200
[R8-bgp]peer 32.32.32.2 as-number 200
[R8-bgp]network 66.66.66.0 24
[R8-bgp]network 32.32.32.0 24
[R8-bgp]quit
[R8]
--------------------------------------------------------------------------------
[R9]bgp 200
[R9-bgp]peer 64.64.64.1 as-number 200
[R9-bgp]peer 63.63.63.2 as-number 200
[R9-bgp]network 64.64.64.0 24
[R9-bgp]network 63.63.63.0 24
[R9-bgp]quit
[R9]
--------------------------------------------------------------------------------
[R10]bgp 200
[R10-bgp]peer 63.63.63.1 as-number 200
[R10-bgp]peer 32.32.32.1 as-number 200
[R10-bgp]network 63.63.63.0 24
[R10-bgp]network 32.32.32.0 24
[R10-bgp]quit
[R10]
```

(4) 任务测试。

```
[R1]display ip routing-table
Route Flags: R - relay, D - download to fib
--------------------------------------------------------------------------------
Routing Tables: Public
         Destinations : 8        Routes : 8
  Destination/Mask   Proto  Pre  Cost      Flags NextHop         Interface
     116.64.64.0/24  Direct 0    0           D   116.64.64.2     GigabitEthernet0/0/0
     116.64.64.2/32  Direct 0    0           D   127.0.0.1       GigabitEthernet0/0/0
   116.64.64.255/32  Direct 0    0           D   127.0.0.1       GigabitEthernet0/0/0
       127.0.0.0/8   Direct 0    0           D   127.0.0.1       InLoopBack0
       127.0.0.1/32  Direct 0    0           D   127.0.0.1       InLoopBack0
```

```
127.255.255.255/32  Direct 0   0        D   127.0.0.1       InLoopBack0
    201.201.201.0/24  IBGP   255  0       RD   116.64.64.1     GigabitEthernet0/0/0
255.255.255.255/32  Direct 0   0        D   127.0.0.1       InLoopBack0
```

由路由表可以看到以下情况。

① BGP 不产生路由，也不发现路由，只是将 BGP 路由表(非 IP 路由表)中路由条目选择性宣告给对等体(可以认为只是转发通告，并不鉴别真伪)。

② 路由器 R2 作为 R1 的 IBGP 对等体，只通告本地 BGP 引入路由(引入方式包括 network 或 import-route)和学习到的 EBGP 路由，不宣告学习到的 IBGP 路由，因此 R1 只能发现(201.201.201.0)网段。

```
[R2]display ip routing-table
Route Flags: R - relay, D - download to fib
------------------------------------------------------------------------------
Routing Tables: Public
         Destinations : 11       Routes : 11
 Destination/Mask   Proto  Pre  Cost      Flags NextHop         Interface
     116.64.64.0/24  Direct 0    0           D   116.64.64.1     GigabitEthernet0/0/0
     116.64.64.1/32  Direct 0    0           D   127.0.0.1       GigabitEthernet0/0/0
   116.64.64.255/32  Direct 0    0           D   127.0.0.1       GigabitEthernet0/0/0
       127.0.0.0/8   Direct 0    0           D   127.0.0.1       InLoopBack0
       127.0.0.1/32  Direct 0    0           D   127.0.0.1       InLoopBack0
 127.255.255.255/32  Direct 0    0           D   127.0.0.1       InLoopBack0
   201.201.201.0/24  Direct 0    0           D   201.201.201.1   GigabitEthernet0/0/1
   201.201.201.1/32  Direct 0    0           D   127.0.0.1       GigabitEthernet0/0/1
 201.201.201.255/32  Direct 0    0           D   127.0.0.1       GigabitEthernet0/0/1
   202.202.202.0/24  IBGP   255  0          RD   201.201.201.2   GigabitEthernet0/0/1
 255.255.255.255/32  Direct 0    0           D   127.0.0.1       InLoopBack0
```

同理，路由器 R3 作为 R2 的 IBGP 对等体，只通告本地引入路由，R2 只能发现(202.202.202.0)网段。

```
[R3]display ip routing-table
Route Flags: R - relay, D - download to fib
------------------------------------------------------------------------------
Routing Tables: Public
         Destinations : 12       Routes : 12
 Destination/Mask   Proto  Pre  Cost      Flags NextHop         Interface
     116.64.64.0/24  IBGP   255  0          RD   201.201.201.1   GigabitEthernet0/0/1
     121.32.32.0/24  IBGP   255  0          RD   202.202.202.1   GigabitEthernet0/0/0
       127.0.0.0/8   Direct 0    0           D   127.0.0.1       InLoopBack0
       127.0.0.1/32  Direct 0    0           D   127.0.0.1       InLoopBack0
 127.255.255.255/32  Direct 0    0           D   127.0.0.1       InLoopBack0
   201.201.201.0/24  Direct 0    0           D   201.201.201.2   GigabitEthernet0/0/1
   201.201.201.2/32  Direct 0    0           D   127.0.0.1       GigabitEthernet0/0/1
 201.201.201.255/32  Direct 0    0           D   127.0.0.1       GigabitEthernet0/0/1
   202.202.202.0/24  Direct 0    0           D   202.202.202.2   GigabitEthernet0/0/0
   202.202.202.2/32  Direct 0    0           D   127.0.0.1       GigabitEthernet0/0/0
 202.202.202.255/32  Direct 0    0           D   127.0.0.1       GigabitEthernet0/0/0
 255.255.255.255/32  Direct 0    0           D   127.0.0.1       InLoopBack0
```

作为IBGP对等体，路由器R2和R4分别向R3通告其本地引入路由，R3只能发现<116.64.64.0>和<121.32.32.0>网段信息。

```
[R4]display ip routing-table
Route Flags: R - relay, D - download to fib
------------------------------------------------------------------------------
Routing Tables: Public
         Destinations : 16        Routes : 16
 Destination/Mask  Proto  Pre  Cost      Flags NextHop         Interface
        0.0.0.0/0  IBGP   255  0          RD  121.32.32.2     GigabitEthernet0/0/1
    63.63.63.0/24  IBGP   255  0          RD  120.16.16.2     GigabitEthernet0/0/1
    64.64.64.0/24  IBGP   255  0          RD  120.16.16.2     GigabitEthernet0/0/1
    65.65.65.0/24  IBGP   255  0          RD  120.16.16.2     GigabitEthernet0/0/1
    66.66.66.0/24  IBGP   255  0          RD  120.16.16.2     GigabitEthernet0/0/1
   120.16.16.0/24  IBGP   255  0          RD  121.32.32.2     GigabitEthernet0/0/1
   121.32.32.0/24  Direct 0    0          D   121.32.32.1     GigabitEthernet0/0/1
   121.32.32.1/32  Direct 0    0          D   127.0.0.1       GigabitEthernet0/0/1
 121.32.32.255/32  Direct 0    0          D   127.0.0.1       GigabitEthernet0/0/1
      127.0.0.0/8  Direct 0    0          D   127.0.0.1       InLoopBack0
     127.0.0.1/32  Direct 0    0          D   127.0.0.1       InLoopBack0
127.255.255.255/32 Direct 0    0          D   127.0.0.1       InLoopBack0
  201.201.201.0/24 IBGP   255  0          RD  202.202.202.2   GigabitEthernet0/0/0
  202.202.202.0/24 Direct 0    0          D   202.202.202.1   GigabitEthernet0/0/0
  202.202.202.1/32 Direct 0    0          D   127.0.0.1       GigabitEthernet0/0/0
202.202.202.255/32 Direct 0    0          D   127.0.0.1       GigabitEthernet0/0/0
255.255.255.255/32 Direct 0    0          D   127.0.0.1       InLoopBack0
```

由以上路由表可以看到以下情况。

① 路由器R5将默认路由下发给R4，R4不再将学习到的默认路由向R3通告。

② 路由器R6作为R5的EBGP对等体，将引入的路由(64.64.64.0)、(65.65.65.0)和通过IBGP学习到的(63.63.63.0)、(66.66.66.0)两个网段(不包含(32.32.32.0))通告给R5。R5作为R4的IBGP对等体，通告本地引入路由<120.16.16.0>和从R6学习到的EBGP路由。

③ 路由器R3将本地引入路由(201.201.201.2)通告给R3，不通告学习到的IBGP路由。

```
[R5]display ip routing-table
Route Flags: R - relay, D - download to fib
------------------------------------------------------------------------------
Routing Tables: Public
         Destinations : 17        Routes : 17
 Destination/Mask  Proto  Pre  Cost      Flags NextHop         Interface
        0.0.0.0/0  Static 60   0          RD  120.16.16.2     Serial2/0/0
    63.63.63.0/24  EBGP   255  0          D   120.16.16.2     Serial2/0/0
    64.64.64.0/24  EBGP   255  0          D   120.16.16.2     Serial2/0/0
    65.65.65.0/24  EBGP   255  0          D   120.16.16.2     Serial2/0/0
    66.66.66.0/24  EBGP   255  0          D   120.16.16.2     Serial2/0/0
   120.16.16.0/24  Direct 0    0          D   120.16.16.1     Serial2/0/0
   120.16.16.1/32  Direct 0    0          D   127.0.0.1       Serial2/0/0
   120.16.16.2/32  Direct 0    0          D   120.16.16.2     Serial2/0/0
 120.16.16.255/32  Direct 0    0          D   127.0.0.1       Serial2/0/0
```

```
      121.32.32.0/24  Direct 0    0        D   121.32.32.2     GigabitEthernet0/0/0
      121.32.32.2/32  Direct 0    0        D   127.0.0.1       GigabitEthernet0/0/0
    121.32.32.255/32  Direct 0    0        D   127.0.0.1       GigabitEthernet0/0/0
         127.0.0.0/8  Direct 0    0        D   127.0.0.1       InLoopBack0
        127.0.0.1/32  Direct 0    0        D   127.0.0.1       InLoopBack0
  127.255.255.255/32  Direct 0    0        D   127.0.0.1       InLoopBack0
    202.202.202.0/24  IBGP   255  0        RD  121.32.32.1     GigabitEthernet0/0/0
  255.255.255.255/32  Direct 0    0        D   127.0.0.1       InLoopBack0
```

可以看到：

① 路由器 R6 作为 R5 的 EBGP 对等体，将引入的路由(64.64.64.0)、(65.65.65.0)和通过 IBGP 学习到的(63.63.63.0)、(66.66.66.0)两个网段(不包含(32.32.32.0))通告给 R5。

② 路由器 R3 向 R5 通告(202.202.202.0)引入路由。

```
[R6]display ip routing-table
Route Flags: R - relay, D - download to fib
------------------------------------------------------------------------------
Routing Tables: Public
         Destinations : 19       Routes : 19
 Destination/Mask    Proto  Pre  Cost      Flags NextHop         Interface
        0.0.0.0/0   Static 60   0           RD   120.16.16.1     Serial2/0/0
     63.63.63.0/24  IBGP   255  0           RD   64.64.64.2      GigabitEthernet0/0/1
     64.64.64.0/24  Direct 0    0           D    64.64.64.1      GigabitEthernet0/0/1
     64.64.64.1/32  Direct 0    0           D    127.0.0.1       GigabitEthernet0/0/1
   64.64.64.255/32  Direct 0    0           D    127.0.0.1       GigabitEthernet0/0/1
     65.65.65.0/24  Direct 0    0           D    65.65.65.1      GigabitEthernet0/0/0
     65.65.65.1/32  Direct 0    0           D    127.0.0.1       GigabitEthernet0/0/0
   65.65.65.255/32  Direct 0    0           D    127.0.0.1       GigabitEthernet0/0/0
     66.66.66.0/24  IBGP   255  0           RD   65.65.65.2      GigabitEthernet0/0/0
    120.16.16.0/24  Direct 0    0           D    120.16.16.2     Serial2/0/0
    120.16.16.1/32  Direct 0    0           D    120.16.16.1     Serial2/0/0
    120.16.16.2/32  Direct 0    0           D    127.0.0.1       Serial2/0/0
  120.16.16.255/32  Direct 0    0           D    127.0.0.1       Serial2/0/0
    121.32.32.0/24  EBGP   255  0           D    120.16.16.1     Serial2/0/0
      127.0.0.0/8   Direct 0    0           D    127.0.0.1       InLoopBack0
      127.0.0.1/32  Direct 0    0           D    127.0.0.1       InLoopBack0
127.255.255.255/32  Direct 0    0           D    127.0.0.1       InLoopBack0
  202.202.202.0/24  EBGP   255  0           D    120.16.16.1     Serial2/0/0
255.255.255.255/32  Direct 0    0           D    127.0.0.1       InLoopBack0
```

可以看到：

① 路由器 R5 作为 R6 的 EBGP 对等体，将引入的路由(121.32.32.0)和通过 IBGP 学习到的(202.202.202.0)网段通告给 R6。

② 路由器 R7 和 R8 向 R6 分别通告引入路由(66.66.66.0)和(63.63.63.0)网段。

```
[R7]display ip routing-table
Route Flags: R - relay, D - download to fib
------------------------------------------------------------------------------
Routing Tables: Public
         Destinations : 16       Routes : 16
Destination/Mask    Proto  Pre  Cost      Flags NextHop         Interface
```

```
         0.0.0.0/0  IBGP    255  0        RD  65.65.65.1      GigabitEthernet0/0/0
      32.32.32.0/24  IBGP    255  0        RD  66.66.66.2      GigabitEthernet0/0/1
      64.64.64.0/24  IBGP    255  0        RD  65.65.65.1      GigabitEthernet0/0/0
      65.65.65.0/24  Direct 0    0        D   65.65.65.2      GigabitEthernet0/0/0
      65.65.65.2/32  Direct 0    0        D   127.0.0.1       GigabitEthernet0/0/0
    65.65.65.255/32  Direct 0    0        D   127.0.0.1       GigabitEthernet0/0/0
      66.66.66.0/24  Direct 0    0        D   66.66.66.1      GigabitEthernet0/0/1
      66.66.66.1/32  Direct 0    0        D   127.0.0.1       GigabitEthernet0/0/1
    66.66.66.255/32  Direct 0    0        D   127.0.0.1       GigabitEthernet0/0/1
      120.16.16.0/24  IBGP    255  0        RD  65.65.65.1      GigabitEthernet0/0/0
      121.32.32.0/24  IBGP    255  0        RD  120.16.16.1     GigabitEthernet0/0/0
        127.0.0.0/8  Direct 0    0        D   127.0.0.1       InLoopBack0
        127.0.0.1/32  Direct 0    0        D   127.0.0.1       InLoopBack0
  127.255.255.255/32  Direct 0    0        D   127.0.0.1       InLoopBack0
    202.202.202.0/24  IBGP    255  0        RD  120.16.16.1     GigabitEthernet0/0/0
  255.255.255.255/32  Direct 0    0        D   127.0.0.1       InLoopBack0
```

可以看到：

① 路由器 R6 将默认路由下发给 R7，R7 不再将学习到的默认路由往 R8 通告。

② 路由器 R5 作为 R6 的 EBGP 对等体，将引入的路由(121.32.32.0)和通过 IBGP 学习到的(202.202.202.0)网段通告给 R6。R6 作为 R7 的 IBGP 对等体，通告本地引入路由(64.64.64.0)、(120.16.16.0)和从 R5 学习到的 EBGP 路由。

③ 路由器 R6 作为 R7 的 IBGP 对等体，不能将学习到的 IBGP 路由(63.63.63.0)通告给 R7。

④ 路由器 R8 作为 R7 的 IBGP 对等体，通告本地引入路由(32.32.32.0)。

```
[R8]dis ip routing-table
Route Flags: R - relay, D - download to fib
------------------------------------------------------------------------------
Routing Tables: Public
         Destinations : 12       Routes : 12
 Destination/Mask    Proto  Pre  Cost     Flags NextHop         Interface
      32.32.32.0/24  Direct 0    0          D   32.32.32.1      GigabitEthernet0/0/1
      32.32.32.1/32  Direct 0    0          D   127.0.0.1       GigabitEthernet0/0/1
    32.32.32.255/32  Direct 0    0          D   127.0.0.1       GigabitEthernet0/0/1
      63.63.63.0/24  IBGP   255  0          RD  32.32.32.2      GigabitEthernet0/0/1
      65.65.65.0/24  IBGP   255  0          RD  66.66.66.1      GigabitEthernet0/0/0
      66.66.66.0/24  Direct 0    0          D   66.66.66.2      GigabitEthernet0/0/0
      66.66.66.2/32  Direct 0    0          D   127.0.0.1       GigabitEthernet0/0/0
    66.66.66.255/32  Direct 0    0          D   127.0.0.1       GigabitEthernet0/0/0
        127.0.0.0/8  Direct 0    0          D   127.0.0.1       InLoopBack0
        127.0.0.1/32  Direct 0    0          D   127.0.0.1       InLoopBack0
 127.255.255.255/32  Direct 0    0          D   127.0.0.1       InLoopBack0
 255.255.255.255/32  Direct 0    0          D   127.0.0.1       InLoopBack0
```

可以看到：

① 路由器 R7 作为 R8 的 IBGP 对等体，通告本地引入路由(65.65.65.0)。

② 路由器 R10 作为 R8 的 IBGP 对等体，通告本地引入路由(63.63.63.0)。

路由器 R9 路由表和 R7 类似，R10 路由表和 R8 类似，这里不再赘述。从路由表可以看到移动和电信组成的 Internet 局部范围不能连通，是错误的配置。

为实现移动和电信全网互通，在不改变物理结构情况下对拓扑重新规划，如图 13-2 所示。

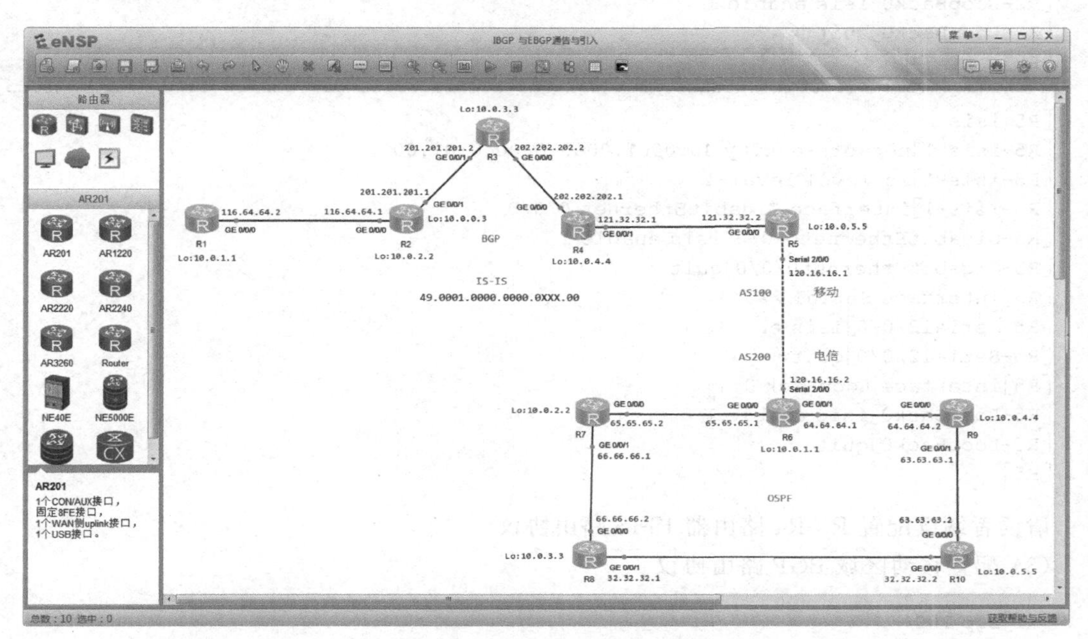

图 13-2 移动电信网络规划图

注：以下是正确配置。

(1) 路由器接口 IP 和系统名称配置。

请读者根据网络拓扑自行配置路由器接口 IP 和系统名称，注意所有接口子网掩码长度均为 24。

(2) 配置移动区域底层 IS-IS 路由实现内网 LoopBack 接口相互连通。

```
[R1]isis 1
[R1-isis-1]network-entity 49.0001.0000.0000.0001.00
[R1-isis-1]is-level level-1
[R1-isis-1]quit
[R1]interface GigabitEthernet 0/0/0
[R1-GigabitEthernet0/0/0]isis enable 1
[R1-GigabitEthernet0/0/0]quit
[R1]interface LoopBack 0
[R1-LoopBack0]isis enable 1
[R1-LoopBack0]quit
[R1]
-----------------------------------------------------------------------------------
[R2]isis 1
[R2-isis-1]network-entity 49.0001.0000.0000.0010.00
[R2-isis-1]is-level level-1
[R2]interface GigabitEthernet 0/0/0
[R2-GigabitEthernet0/0/0]isis enable 1
[R2-GigabitEthernet0/0/0]quit
[R2]interface GigabitEthernet 0/0/1
[R2-GigabitEthernet0/0/1]isis enable 1
```

```
[R2-GigabitEthernet0/0/1]quit
[R2]interface LoopBack 0
[R2-LoopBack0]isis enable 1
[R2-LoopBack0]quit
[R2]
--------------------------------------------------------------------------------
[R5]isis 1
[R5-isis-1]network-entity 49.0001.0000.0000.0101.00
[R5-isis-1]is-level level-1
[R5-isis-1]interface GigabitEthernet 0/0/0
[R5-GigabitEthernet0/0/0]isis enable 1
[R5-GigabitEthernet0/0/0]quit
[R5]interface Serial 2/0/0
[R5-Serial2/0/0]isis enable 1
[R5-Serial2/0/0]quit
[R5]interface LoopBack 0
[R5-LoopBack0]isis enable 1
[R5-LoopBack0]quit
[R5]
```

请读者继续配置 R3-R4 路由器 IS-IS 路由协议。

(3) 配置移动区域 BGP 路由协议。

```
[R1]bgp 100
[R1-bgp]peer 10.0.5.5 as-number 100  //与 R2 的 LooBack 接口建立 IBGP 邻居。IBGP 一般通
                                       过 LooBack 虚拟接口建立邻居关系,避免物理接口 IP
                                       更换或者线路故障导致邻接关系中断
[R1-bgp]peer 10.0.5.5 connect-interface LoopBack 0  //手动指定建立邻居的接口为本地
                                                     LoopBack 0
```

注意:

① 配置 BGP 对等体(邻居)时使用的 IP 地址,应互为 BGP 报文的源 IP 地址和目的 IP 地址。默认情况下,BGP 使用去往对等体路由器的出口 IP 作为建立 BGP 邻居的源 IP 地址。此 R1 采用源 IP 地址(116.64.64.2)与对等体 IP 地址(10.0.5.5)建立 IBGP 邻居,而 R5 采用源 IP 地址(121.32.32.2)与对等体 IP 地址(10.0.1.1)建立 IBGP 邻居关系,两者非互为源 IP 地址和目的 IP 地址,导致 R1 与 R5 无法建立 IBGP 对等体。此时需手动指定建立 IBGP 对等体所使用的源接口 IP 地址,即 R1 采用 LoopBack 0(10.0.1.1)作为源 IP 地址,与对等体 IP 地址(10.0.5.5)建立邻居;同样 R5 将 LoopBack 0(10.0.5.5)作为源 IP 地址,与 IP 地址(10.0.1.1)建立对等体,两者互为 IBGP 报文的源 IP 地址和目的 IP 地址。

② 建立对等体的目的在于相互交换 BGP 路由分组。路由器 R1 没必要与 R2、R3 和 R4 建立对等体,因为 R2、R3 和 R4 不会将学习到的 IBGP 路由宣告给 R1,直接与 R5 建立对等体交换 BGP 路由即可。

③ BGP 没有邻居概念(邻居是物理范畴),BGP 称为对等体(逻辑范畴)。但现在已模糊邻居和对等体概念,能理解即可。

```
[R1-bgp]peer 10.0.5.5 enable  //允许对等体(处于邻居关系的路由器称为对等体)10.0.5.5 与
                                R1 交换 IBGP 路由信息。如默认允许,本行命令可以不输入
[R1-bgp]quit
[R1]
--------------------------------------------------------------------------------
```

```
[R2]bgp 100
[R2-bgp]peer 10.0.5.5 as-number 100
[R2-bgp]peer 10.0.5.5 connect-interface LoopBack 0
[R2-bgp]quit
[R2]
------------------------------------------------------------------------
[R3]bgp 100
[R3-bgp]peer 10.0.5.5 as-number 100
[R3-bgp]peer 10.0.5.5 connect-interface LoopBack 0
[R3-bgp]quit
[R3]
------------------------------------------------------------------------
[R4]bgp 100
[R4-bgp]peer 10.0.5.5 as-number 100
[R4-bgp]peer 10.0.5.5 connect-interface LoopBack 0
[R4-bgp]quit
[R4]
------------------------------------------------------------------------
[R5]bgp 100
[R5-bgp]peer 10.0.1.1 as-number 100
[R5-bgp]peer 10.0.1.1 connect-interface LoopBack 0
[R5-bgp]peer 10.0.2.2 as-number 100
[R5-bgp]peer 10.0.2.2 connect-interface LoopBack 0
[R5-bgp]peer 10.0.3.3 as-number 100
[R5-bgp]peer 10.0.3.3 connect-interface LoopBack 0
[R5-bgp]peer 10.0.4.4 as-number 100
[R5-bgp]peer 10.0.4.4 connect-interface LoopBack 0
[R5-bgp]peer 120.16.16.2 as-number 200  //不同AS之间,EBGP建议采用物理接口建立邻居
                                          (EBGP邻居之间在发送BGP报文时,TTL值默认为
                                          1,所以EBGP默认要求邻居之间必须物理直连)
[R5-bgp]quit
[R5]
```

(4) 配置电信区域底层 OSPF 路由实现内网 LoopBack 接口相互连通。

```
[R6]ospf 1
[R6-ospf-1]area 0
[R6-ospf-1-area-0.0.0.0]network 65.65.65.0 0.0.0.255
[R6-ospf-1-area-0.0.0.0]network 64.64.64.0 0.0.0.255
[R6-ospf-1-area-0.0.0.0]network 120.16.16.0 0.0.0.255
[R6-ospf-1-area-0.0.0.0]network 10.0.1.1 0.0.0.0
[R6-ospf-1-area-0.0.0.0]quit
[R6-ospf-1]quit
[R6]
------------------------------------------------------------------------
[R7]ospf 1
[R7-ospf-1]area 0
[R7-ospf-1-area-0.0.0.0]network 65.65.65.0 0.0.0.255
[R7-ospf-1-area-0.0.0.0]network 66.66.66.0 0.0.0.255
[R7-ospf-1-area-0.0.0.0]network 10.0.2.2 0.0.0.0
[R7-ospf-1-area-0.0.0.0]quit
[R7-ospf-1]quit
[R7]
```

请读者继续配置 R8～R10 路由器 OSPF 路由协议。

（5）配置电信区域 BGP 路由协议。

```
[R6]bgp 200
[R6-bgp]peer 120.16.16.1 as 100      //在不同 AS 之间，EBGP 建议采用物理接口建立邻居
[R6-bgp]peer 10.0.2.2 as-number 200
[R6-bgp]peer 10.0.2.2 connect-interface LoopBack 0
[R6-bgp]peer 10.0.3.3 as-number 200
[R6-bgp]peer 10.0.3.3 connect-interface LoopBack 0
[R6-bgp]peer 10.0.4.4 as-number 200
[R6-bgp]peer 10.0.4.4 connect-interface LoopBack 0
[R6-bgp]peer 10.0.5.5 as-number 200
[R6-bgp]peer 10.0.5.5 connect-interface LoopBack 0
[R6-bgp]quit
[R6]
-----------------------------------------------------------------------------------
[R7]bgp 200
[R7-bgp]peer 10.0.1.1 as-number 200
[R7-bgp]peer 10.0.1.1 connect-interface LoopBack 0
[R7-bgp]quit
[R7]
-----------------------------------------------------------------------------------
[R8]bgp 200
[R8-bgp]peer 10.0.1.1 as-number 200
[R8-bgp]peer 10.0.1.1 connect-interface LoopBack 0
[R8-bgp]quit
[R8]
-----------------------------------------------------------------------------------
[R9]bgp 200
[R9-bgp]peer 10.0.1.1 as-number 200
[R9-bgp]peer 10.0.1.1 connect-interface LoopBack 0
[R9-bgp]quit
[R9]
-----------------------------------------------------------------------------------
[R10]bgp 200
[R10-bgp]peer 10.0.1.1 as-number 200
[R10-bgp]peer 10.0.1.1 connect-interface LoopBack 0
[R10-bgp]quit
[R10]
```

（6）在路由器 R5 和 R6 上引入 IGP 路由。

```
[R5]bgp 100
[R5-bgp]import-route isis 1
[R5-bgp]quit
[R5]
-----------------------------------------------------------------------------------
[R6]bgp 200
[R6-bgp]import-route ospf 1
[R6-bgp]quit
[R6]
```

(7) 在路由器 R5 和 R6 上查看 BGP 对等体建立情况。

```
[R6]display bgp peer
 BGP local router ID : 65.65.65.1
 Local AS number : 200
 Total number of peers : 5          Peers in established state : 5
  Peer        V   AS  MsgRcvd  MsgSent  OutQ   Up/Down   State Pref   Rcv
  10.0.2.2    4   200       8       11     0  00:06:34  Established    0
  10.0.3.3    4   200       7        8     0  00:05:33  Established    0
  10.0.4.4    4   200       6        7     0  00:04:50  Established    0
  10.0.5.5    4   200       6        8     0  00:04:04  Established    0
  120.16.16.1 4   100      13       13     0  00:11:33  Established   10
```

路由器 R6 已经和 R7、R8、R9、R10 和 R5 成功建立对等体,当前状态为 Established。

```
[R5]display bgp peer
BGP local router ID : 121.32.32.2
Local AS number : 100
Total number of peers : 5            Peers in established state : 2
  Peer        V   AS  MsgRcvd  MsgSent  OutQ   Up/Down   State Pref   Rcv
  10.0.1.1    4   100       9       20     0  00:07:51  Established    0
  10.0.2.2    4   100       0        1     0  00:08:55  Connect        0
  10.0.3.3    4   100       0        0     0  00:05:16  Connect        0
  10.0.4.4    4   100       0        0     0  00:05:07  Connect        0
  120.16.16.2 4   200      23       18     0  00:08:23  Established   11
```

发现路由器 R5 无法与路由器 R2、R3 和 R4 建立对等体,处于正在连接 Connect 状态,说明 LoopBack 接口无法连通。可检查 R5 路由表如下:

```
[R5]display ip routing-table
Route Flags: R - relay, D - download to fib
------------------------------------------------------------------------------
Routing Tables: Public
         Destinations : 29        Routes : 29
  Destination/Mask  Proto  Pre  Cost     Flags NextHop         Interface
       10.0.1.0/24  ISIS- L15   40         D   121.32.32.1     GigabitEthernet0/0/0
       10.0.2.0/24  ISIS- L15   30         D   121.32.32.1     GigabitEthernet0/0/0
       10.0.2.2/32  EBGP   255  1          D   120.16.16.2     Serial2/0/0
       10.0.3.0/24  ISIS- L15   20         D   121.32.32.1     GigabitEthernet0/0/0
       10.0.3.3/32  EBGP   255  2          D   120.16.16.2     Serial2/0/0
       10.0.4.0/24  ISIS- L15   10         D   121.32.32.1     GigabitEthernet0/0/0
       10.0.4.4/32  EBGP   255  1          D   120.16.16.2     Serial2/0/0
       10.0.5.0/24  Direct 0    0          D   10.0.5.5        LoopBack0
       10.0.5.5/32  Direct 0    0          D   127.0.0.1       LoopBack0
     10.0.5.255/32  Direct 0    0          D   127.0.0.1       LoopBack0
     32.32.32.0/24  EBGP   255  3          D   120.16.16.2     Serial2/0/0
     63.63.63.0/24  EBGP   255  2          D   120.16.16.2     Serial2/0/0
     64.64.64.0/24  EBGP   255  0          D   120.16.16.2     Serial2/0/0
     65.65.65.0/24  EBGP   255  0          D   120.16.16.2     Serial2/0/0
     66.66.66.0/24  EBGP   255  2          D   120.16.16.2     Serial2/0/0
    116.64.64.0/24  ISIS- L15   40         D   121.32.32.1     GigabitEthernet0/0/0
    120.16.16.0/24  Direct 0    0          D   120.16.16.1     Serial2/0/0
    120.16.16.1/32  Direct 0    0          D   127.0.0.1       Serial2/0/0
```

```
   120.16.16.2/32  Direct 0    0         D  120.16.16.2     Serial2/0/0
 120.16.16.255/32  Direct 0    0         D  127.0.0.1       Serial2/0/0
    121.32.32.0/24  Direct 0    0         D  121.32.32.2     GigabitEthernet0/0/0
    121.32.32.2/32  Direct 0    0         D  127.0.0.1       GigabitEthernet0/0/0
  121.32.32.255/32  Direct 0    0         D  127.0.0.1       GigabitEthernet0/0/0
       127.0.0.0/8  Direct 0    0         D  127.0.0.1       InLoopBack0
      127.0.0.1/32  Direct 0    0         D  127.0.0.1       InLoopBack0
127.255.255.255/32  Direct 0    0         D  127.0.0.1       InLoopBack0
  201.201.201.0/24  ISIS- L15   30        D  121.32.32.1     GigabitEthernet0/0/0
  202.202.202.0/24  ISIS- L15   20        D  121.32.32.1     GigabitEthernet0/0/0
255.255.255.255/32  Direct 0    0         D  127.0.0.1       InLoopBack0
```

由于路由器R6引入OSPF路由，将电信<10.0.0.0>内网向R5通告，如R5去往10.0.2.2存在两条路径，分别是<10.0.2.0/24>(ISIS)和<10.0.2.2/32>(EBGP)。IP路由表采用精确匹配优先原则，去往"10.0.2.2/32""10.0.3.3/32"和"10.0.4.4/32"通过R6转发，导致LoopBack接口无法连通。以<10.0.2.2/32>为例验证如下：

```
[R5]tracert 10.0.2.2
traceroute to 10.0.2.2(10.0.2.2), max hops: 30 ,packet length: 40,press CTRL_C
to break
1 120.16.16.2 20 ms  30 ms  20 ms
```

这将导致移动和电信网络无法连通，测试如下：

```
[R1]ping 32.32.32.2          //R1 ping R10
  PING 32.32.32.2: 56  data bytes, press CTRL_C to break
    Request time out
    Request time out
    Request time out
    Request time out
    Request time out
  --- 32.32.32.2 ping statistics ---
    5 packet(s) transmitted
    0 packet(s) received
    100.00%  packet loss
```

查看R6路由表如下：

```
[R6]display ip routing-table
Route Flags: R - relay, D - download to fib
------------------------------------------------------------------------------
Routing Tables: Public
         Destinations : 32       Routes : 33
  Destination/Mask  Proto  Pre  Cost     Flags NextHop         Interface
        10.0.1.0/24  Direct 0    0         D  10.0.1.1        LoopBack0
        10.0.1.1/32  Direct 0    0         D  127.0.0.1       LoopBack0
      10.0.1.255/32  Direct 0    0         D  127.0.0.1       LoopBack0
        10.0.2.0/24  EBGP   255  30        D  120.16.16.1     Serial2/0/0
        10.0.2.2/32  OSPF   10   1         D  65.65.65.2      GigabitEthernet0/0/0
        10.0.3.0/24  EBGP   255  20        D  120.16.16.1     Serial2/0/0
        10.0.3.3/32  OSPF   10   2         D  65.65.65.2      GigabitEthernet0/0/0
        10.0.4.0/24  EBGP   255  10        D  120.16.16.1     Serial2/0/0
        10.0.4.4/32  OSPF   10   1         D  64.64.64.2      GigabitEthernet0/0/1
```

```
        10.0.5.0/24  EBGP    255  0          D   120.16.16.1    Serial2/0/0
        10.0.5.5/32  OSPF    10   2          D   64.64.64.2     GigabitEthernet0/0/1
      32.32.32.0/24  OSPF    10   3          D   65.65.65.2     GigabitEthernet0/0/0
                     OSPF    10   3          D   64.64.64.2     GigabitEthernet0/0/1
      63.63.63.0/24  OSPF    10   2          D   64.64.64.2     GigabitEthernet0/0/1
      64.64.64.0/24  Direct 0     0          D   64.64.64.1     GigabitEthernet0/0/1
      64.64.64.1/32  Direct 0     0          D   127.0.0.1      GigabitEthernet0/0/1
    64.64.64.255/32  Direct 0     0          D   127.0.0.1      GigabitEthernet0/0/1
      65.65.65.0/24  Direct 0     0          D   65.65.65.1     GigabitEthernet0/0/0
      65.65.65.1/32  Direct 0     0          D   127.0.0.1      GigabitEthernet0/0/0
    65.65.65.255/32  Direct 0     0          D   127.0.0.1      GigabitEthernet0/0/0
      66.66.66.0/24  OSPF    10   2          D   65.65.65.2     GigabitEthernet0/0/0
     116.64.64.0/24  EBGP    255  40         D   120.16.16.1    Serial2/0/0
     120.16.16.0/24  Direct 0     0          D   120.16.16.2    Serial2/0/0
     120.16.16.1/32  Direct 0     0          D   120.16.16.1    Serial2/0/0
     120.16.16.2/32  Direct 0     0          D   127.0.0.1      Serial2/0/0
   120.16.16.255/32  Direct 0     0          D   127.0.0.1      Serial2/0/0
     121.32.32.0/24  EBGP    255  0          D   120.16.16.1    Serial2/0/0
        127.0.0.0/8  Direct 0     0          D   127.0.0.1      InLoopBack0
       127.0.0.1/32  Direct 0     0          D   127.0.0.1      InLoopBack0
127.255.255.255/32   Direct 0     0          D   127.0.0.1      InLoopBack0
   201.201.201.0/24  EBGP    255  30         D   120.16.16.1    Serial2/0/0
   202.202.202.0/24  EBGP    255  20         D   120.16.16.1    Serial2/0/0
255.255.255.255/32   Direct 0     0          D   127.0.0.1      InLoopBack0
```

同样，路由器 R6 去往<10.0.0.0>网段同样存在两条路径，分别是 EBGP 和 OSPF。为避免电信和移动相互通告<10.0.0.0>私用网段，可在 BGP 通告中采用路由信息过滤。

(8) 过滤电信和移动相互通告<10.0.0.0>私用网段。

```
[R5]ip ip-prefix deny_net_10 deny 10.0.0.0 8 greater-equal 8   //大于或等于 8 的网络掩码
[R5]ip ip-prefix deny_net_10 index 20 permit 0.0.0.0 0 less-equal 32
//无论是 ACL 还是 IP Prefix 过滤，默认都是配置 deny all，所以在拒绝需要过滤的路由条目后，最
  后还要配置一条 permit 命令让其他路由通过
[R5]bgp 100
[R5-bgp]filter-policy ip-prefix deny_net_10 export                //注意不是 import
[R5-bgp]quit
[R5]
------------------------------------------------------------------------------------------
[R6]ip ip-prefix deny_net_10 deny 10.0.0.0 8 greater-equal 8
[R6]ip ip-prefix deny_net_10 index 20 permit 0.0.0.0 0 less-equal 32
[R6]bgp 200
[R6-bgp]filter-policy ip-prefix deny_net_10 export
[R6-bgp]quit
[R6]
```

2. 任务验证

(1) 在路由器 R5 和 R6 上重新查看 IP 路由表。

```
[R5]display ip routing-table
Route Flags: R - relay, D - download to fib
```

```
-------------------------------------------------------------------------------
Routing Tables: Public
         Destinations : 26       Routes : 26
  Destination/Mask   Proto  Pre  Cost      Flags NextHop         Interface
        10.0.1.0/24  ISIS- L15   40          D   121.32.32.1     GigabitEthernet0/0/0
        10.0.2.0/24  ISIS- L15   30          D   121.32.32.1     GigabitEthernet0/0/0
        10.0.3.0/24  ISIS- L15   20          D   121.32.32.1     GigabitEthernet0/0/0
        10.0.4.0/24  ISIS- L15   10          D   121.32.32.1     GigabitEthernet0/0/0
        10.0.5.0/24  Direct 0    0           D   10.0.5.5        LoopBack0
        10.0.5.5/32  Direct 0    0           D   127.0.0.1       LoopBack0
      10.0.5.255/32  Direct 0    0           D   127.0.0.1       LoopBack0
      32.32.32.0/24  EBGP   255  3           D   120.16.16.2     Serial2/0/0
      63.63.63.0/24  EBGP   255  2           D   120.16.16.2     Serial2/0/0
      64.64.64.0/24  EBGP   255  0           D   120.16.16.2     Serial2/0/0
      65.65.65.0/24  EBGP   255  0           D   120.16.16.2     Serial2/0/0
      66.66.66.0/24  EBGP   255  2           D   120.16.16.2     Serial2/0/0
     116.64.64.0/24  ISIS- L15   40          D   121.32.32.1     GigabitEthernet0/0/0
     120.16.16.0/24  Direct 0    0           D   120.16.16.1     Serial2/0/0
     120.16.16.1/32  Direct 0    0           D   127.0.0.1       Serial2/0/0
     120.16.16.2/32  Direct 0    0           D   120.16.16.2     Serial2/0/0
   120.16.16.255/32  Direct 0    0           D   127.0.0.1       Serial2/0/0
     121.32.32.0/24  Direct 0    0           D   121.32.32.2     GigabitEthernet0/0/0
     121.32.32.2/32  Direct 0    0           D   127.0.0.1       GigabitEthernet0/0/0
   121.32.32.255/32  Direct 0    0           D   127.0.0.1       GigabitEthernet0/0/0
        127.0.0.0/8  Direct 0    0           D   127.0.0.1       InLoopBack0
       127.0.0.1/32  Direct 0    0           D   127.0.0.1       InLoopBack0
 127.255.255.255/32  Direct 0    0           D   127.0.0.1       InLoopBack0
   201.201.201.0/24  ISIS- L15   30          D   121.32.32.1     GigabitEthernet0/0/0
   202.202.202.0/24  ISIS- L15   20          D   121.32.32.1     GigabitEthernet0/0/0
 255.255.255.255/32  Direct 0    0           D   127.0.0.1       InLoopBack0
```

可以看到，R5不再拥有去往<10.0.0.0>网段的EBGP路由。

```
[R6]display ip routing-table
Route Flags: R - relay, D - download to fib
-------------------------------------------------------------------------------
Routing Tables: Public
         Destinations : 28       Routes : 29
  Destination/Mask   Proto  Pre  Cost      Flags NextHop         Interface
        10.0.1.0/24  Direct 0    0           D   10.0.1.1        LoopBack0
        10.0.1.1/32  Direct 0    0           D   127.0.0.1       LoopBack0
      10.0.1.255/32  Direct 0    0           D   127.0.0.1       LoopBack0
        10.0.2.2/32  OSPF   10   1           D   65.65.65.2      GigabitEthernet0/0/0
        10.0.3.3/32  OSPF   10   2           D   65.65.65.2      GigabitEthernet0/0/0
        10.0.4.4/32  OSPF   10   1           D   64.64.64.2      GigabitEthernet0/0/1
        10.0.5.5/32  OSPF   10   2           D   64.64.64.2      GigabitEthernet0/0/1
      32.32.32.0/24  OSPF   10   3           D   65.65.65.2      GigabitEthernet0/0/0
                     OSPF   10   3           D   64.64.64.2      GigabitEthernet0/0/1
      63.63.63.0/24  OSPF   10   2           D   64.64.64.2      GigabitEthernet0/0/1
      64.64.64.0/24  Direct 0    0           D   64.64.64.1      GigabitEthernet0/0/1
      64.64.64.1/32  Direct 0    0           D   127.0.0.1       GigabitEthernet0/0/1
    64.64.64.255/32  Direct 0    0           D   127.0.0.1       GigabitEthernet0/0/1
```

```
     65.65.65.0/24  Direct 0    0        D   65.65.65.1      GigabitEthernet0/0/0
     65.65.65.1/32  Direct 0    0        D   127.0.0.1       GigabitEthernet0/0/0
   65.65.65.255/32  Direct 0    0        D   127.0.0.1       GigabitEthernet0/0/0
     66.66.66.0/24  OSPF   10   2        D   65.65.65.2      GigabitEthernet0/0/0
    116.64.64.0/24  EBGP   255  40       D   120.16.16.1     Serial2/0/0
    120.16.16.0/24  Direct 0    0        D   120.16.16.2     Serial2/0/0
    120.16.16.1/32  Direct 0    0        D   120.16.16.1     Serial2/0/0
    120.16.16.2/32  Direct 0    0        D   127.0.0.1       Serial2/0/0
  120.16.16.255/32  Direct 0    0        D   127.0.0.1       Serial2/0/0
    121.32.32.0/24  EBGP   255  0        D   120.16.16.1     Serial2/0/0
       127.0.0.0/8  Direct 0    0        D   127.0.0.1       InLoopBack0
      127.0.0.1/32  Direct 0    0        D   127.0.0.1       InLoopBack0
127.255.255.255/32  Direct 0    0        D   127.0.0.1       InLoopBack0
   201.201.201.0/24 EBGP   255  30       D   120.16.16.1     Serial2/0/0
   202.202.202.0/24 EBGP   255  20       D   120.16.16.1     Serial2/0/0
255.255.255.255/32  Direct 0    0        D   127.0.0.1       InLoopBack0
```

同样，路由器 R6 也不再拥有去往<10.0.0.0>网段的 EBGP 路由。

(2) 在路由器 R5 上重新查看 BGP 对等体建立情况。

```
[R5]display bgp peer
BGP local router ID : 121.32.32.2
Local AS number : 100
Total number of peers : 5          Peers in established state : 5
  Peer            V          AS  MsgRcvd  MsgSent  OutQ  Up/Down     State Pref  Rcv

  10.0.1.1        4         100       66       79     0  01:04:21  Established      0
  10.0.2.2        4         100       18       30     0  00:16:36  Established      0
  10.0.3.3        4         100       18       29     0  00:16:32  Established      0
  10.0.4.4        4         100       18       29     0  00:16:24  Established      0
  120.16.16.2     4         200       80       75     0  01:04:53  Established      6
```

路由器 R5 已经和 R1、R2、R3、R4 和 R6 成功建立对等体，当前状态为 Established。

(3) 连通性验证。

移动和电信全网互通。以路由器 R1 与 R10 为例，测试如下：

```
[R1]ping 32.32.32.2                      //R1 ping R10
  PING 32.32.32.2: 56 data bytes, press CTRL_C to break
    Reply from 32.32.32.2: bytes=56 Sequence=1 ttl=249 time=70 ms
    Reply from 32.32.32.2: bytes=56 Sequence=2 ttl=249 time=50 ms
    Reply from 32.32.32.2: bytes=56 Sequence=3 ttl=249 time=60 ms
    Reply from 32.32.32.2: bytes=56 Sequence=4 ttl=249 time=60 ms
    Reply from 32.32.32.2: bytes=56 Sequence=5 ttl=249 time=40 ms
  --- 32.32.32.2 ping statistics ---
    5 packet(s) transmitted
    5 packet(s) received
    0.00% packet loss
    round-trip min/avg/max = 40/56/70 ms
```

【任务总结】

(1) IBGP→IBGP。只通告本地 BGP 引入路由(引入方式包括 network 或 import-route)和学习到的 EBGP 路由,不宣告学习到的 IBGP 路由。

(2) EBGP→EBGP。通告本地引入路由、学习到的 BGP 路由(包括 IBGP 和 EBGP 路由)。

(3) 当网络边缘路由器向内部路由器通告时,将 EBGP 转换为 IBGP 路由条目并进行通告。

工作任务十四

BGP 综合任务

【工作目的】

掌握公网多个 AS 之间 BGP 组建过程。

【工作任务】

电信、移动和联通三大运营商组成公网拓扑。其中电信和联通内网采用 IS-IS 路由组建，移动内网采用 OSPF 路由组建，运营商之间通过边界路由器 R3、R5 和 R11 连接在一起。需要对公网路由器配置 BGP 协议，为公司 A、公司 B 和公司 C 提供网络接入服务，具体任务如下。

(1) 运营商内部路由器分别与其边界路由器建立 IBGP 对等体，以获得其他运营商路由条目。

(2) 运营商边界路由器之间相互建立 EBGP 对等体，并通过 BGP 协议引入内网路由，将其发布给对等体。从而实现运营商之间公网互通。

(3) 禁止运营商对外宣告<10.0.0.0>私用路由条目。

(4) 配置 Easy-IP，主机 1～主机 6 可以访问公网任意网段。

【设备器材】

三层交换机(S5700)3 台，路由器(AR1220)15 台，部分需添加 1GEC 千兆接口模块或 2SA 串口模块。

【环境拓扑】

环境拓扑如图 14-1 所示。

【工作过程】

1. 基本配置

(1) 路由器接口 IP 和系统名称配置。

请读者根据网络拓扑自行配置路由器接口 IP 和系统名称，注意所有接口子网掩码长度均为 24。

(2) 电信网络 BGP 配置。

① 配置电信内网底层 IS-IS 路由，实现 LoopBack 接口之间连通。下面以路由器 R1 和 R2 为例配置 IS-IS 路由。

```
[R1]isis 1
[R1-isis-1]network-entity 10.0001.0000.0000.0001.00
```

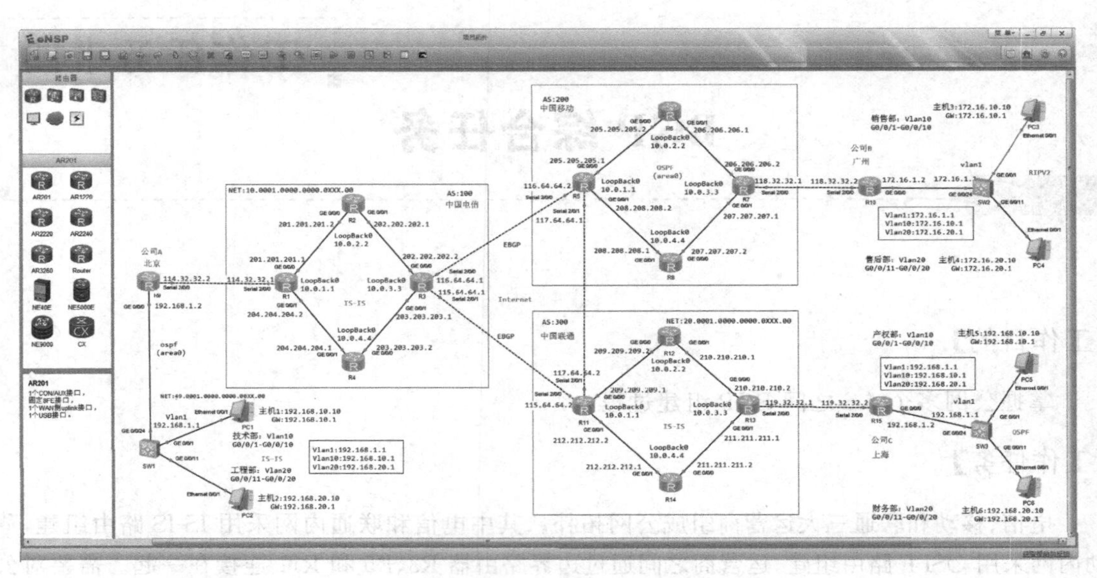

图 14-1 环境拓扑

```
[R1-isis-1]is-level level-1
[R1-isis-1]quit
[R1]interface GigabitEthernet 0/0/0
[R1-GigabitEthernet0/0/0]isis enable 1
[R1-GigabitEthernet0/0/0]quit
[R1]interface GigabitEthernet 0/0/1
[R1-GigabitEthernet0/0/1]isis enable 1
[R1-GigabitEthernet0/0/1]quit
[R1]interface Serial 2/0/0
[R1-Serial2/0/0]isis enable 1
[R1-Serial2/0/0]quit
[R1]interface LoopBack 0
[R1-LoopBack0]isis enable 1
[R1-LoopBack0]quit
[R1]
```

```
[R2]isis 1
[R2-isis-1]network-entity 10.0001.0000.0000.0010.00
[R2-isis-1]is-level level-1
[R2-isis-1]quit
[R2]interface GigabitEthernet 0/0/0
[R2-GigabitEthernet0/0/0]isis enable 1
[R2-GigabitEthernet0/0/0]quit
[R2]interface GigabitEthernet 0/0/1
[R2-GigabitEthernet0/0/1]isis enable 1
[R2-GigabitEthernet0/0/1]quit
[R2]interface LoopBack 0
[R2-LoopBack0]isis enable 1
[R2-LoopBack0]quit
[R2]
```

请读者继续配置路由器 R3 和 R4 的 IS-IS 路由，注意区域内网络实体名称不能冲突，否则无法建立 IS-IS 邻居。

• R3 配置参数。

网络实体名称：10.0001.0000.0000.0011.00。

IS-IS 通告接口：GE0/0/0、GE0/0/1、S2/0/0、S2/0/1、LoopBack 0。

• R4 配置参数。

网络实体名称：10.0001.0000.0000.0100.00。

IS-IS 通告接口：GE0/0/0、GE0/0/1、LoopBack 0。

② 配置 BGP 协议实现电信和移动、联通网络互联。

路由器 R3 连接移动和联通网络，属于区域边界路由器。路由器 R1、R2 和 R4 分别与 R3 建立 IBGP 邻接关系，以学习到其他运营商内网路由。

```
[R1]bgp 100
[R1-bgp]peer 10.0.3.3 as-number 100
[R1-bgp]peer 10.0.3.3 connect-interface LoopBack0
[R1-bgp]quit
[R1]
------------------------------------------------------------------------------------
[R2]bgp 100
[R2-bgp]peer 10.0.3.3 as-number 100
[R2-bgp]peer 10.0.3.3 connect-interface LoopBack0
[R2-bgp]quit
[R2]
------------------------------------------------------------------------------------
[R4]bgp 100
[R4-bgp]peer 10.0.3.3 as-number 100
[R4-bgp]peer 10.0.3.3 connect-interface LoopBack0
[R4-bgp]quit
[R4]
------------------------------------------------------------------------------------
[R3]bgp 100
[R3-bgp]peer 10.0.1.1 as-number 100
[R3-bgp]peer 10.0.1.1 connect-interface LoopBack0
[R3-bgp]peer 10.0.2.2 as-number 100
[R3-bgp]peer 10.0.2.2 connect-interface LoopBack 0
[R3-bgp]peer 10.0.4.4 as-number 100
[R3-bgp]peer 10.0.4.4 connect-interface LoopBack 0
[R3-bgp]peer 116.64.64.2 as-number 200
[R3-bgp]peer 115.64.64.2 as-number 300
[R3-bgp]import-route isis 1  //在 BGP 中导入本地 ISIS 路由，目的是将其以 EBGP 方式宣告给
                               对等体 R5 和 R11，让其学习到电信私网路由，并将路由信息向其
                               对等体 R6、R7 和 R8，R12、R13 和 R14 宣告

[R3-bgp]quit
[R3]
```

③ 过滤中国电信对外通告私网<10.0.0.0>私用网段。

```
[R3]ip ip-prefix deny_net_10 deny 10.0.0.0 8 greater-equal 8
[R3]ip ip-prefix deny_net_10 index 20 permit 0.0.0.0 0 less-equal 32
[R3]bgp 100
```

```
[R3-bgp]filter-policy ip-prefix deny_net_10 export
[R3-bgp]quit
[R3]
```

(3) 移动网络 BGP 配置。

① 配置移动内网底层 OSPF 路由，实现 LooBack 接口相互连通。下面以路由器 R5 和 R6 为例配置 OSPF 路由。

```
[R5]ospf 1
[R5-ospf-1]area 0
[R5-ospf-1-area-0.0.0.0]network 117.64.64.0 0.0.0.255
[R5-ospf-1-area-0.0.0.0]network 116.64.64.0 0.0.0.255
[R5-ospf-1-area-0.0.0.0]network 205.205.205.0 0.0.0.255
[R5-ospf-1-area-0.0.0.0]network 208.208.208.0 0.0.0.255
[R5-ospf-1-area-0.0.0.0]network 10.0.1.1 0.0.0.0
[R5-ospf-1-area-0.0.0.0]quit
[R5-ospf-1]quit
[R5]
------------------------------------------------------------------------------------------
[R6]ospf 1
[R6-ospf-1]area 0
[R6-ospf-1-area-0.0.0.0]network 205.205.205.0 0.0.0.255
[R6-ospf-1-area-0.0.0.0]network 206.206.206.0 0.0.0.255
[R6-ospf-1-area-0.0.0.0]network 10.0.2.2 0.0.0.0
[R6-ospf-1-area-0.0.0.0]quit
[R6-ospf-1]quit
[R6]
```

请读者继续配置路由器 R7 和 R8 的 OSPF 路由，宣告所有接口直连网段信息。

② 配置 BGP 协议实现移动和电信、联通网络互联。

R5 连接电信和联通网络，属于区域边界路由器。路由器 R6、R7 和 R8 分别与 R5 建立 IBGP 邻接关系，以学习到其他运营商内网路由。

```
[R5-bgp]peer 10.0.2.2 as-number 200
[R5-bgp]peer 10.0.2.2 connect-interface LoopBack0
[R5-bgp]peer 10.0.3.3 as-number 200
[R5-bgp]peer 10.0.3.3 connect-interface LoopBack 0
[R5-bgp]peer 10.0.4.4 as-number 200
[R5-bgp]peer 10.0.4.4 connect-interface LoopBack 0
[R5-bgp]peer 116.64.64.1 as-number 100
[R5-bgp]peer 117.64.64.2 as-number 300
[R5-bgp]import-route ospf 1
[R5-bgp]quit
[R5]
------------------------------------------------------------------------------------------
[R6]bgp 200
[R6-bgp]peer 10.0.1.1 as-number 200
[R6-bgp]peer 10.0.1.1 connect-interface LoopBack 0
[R6-bgp]quit
[R6]
------------------------------------------------------------------------------------------
```

```
[R7]bgp 200
[R7-bgp]peer 10.0.1.1 as-number 200
[R7-bgp]peer 10.0.1.1 connect-interface LoopBack 0
[R7-bgp]quit
[R7]
--------------------------------------------------------------------------------
[R8]bgp 200
[R8-bgp]peer 10.0.1.1 as-number 200
[R8-bgp]peer 10.0.1.1 connect-interface LoopBack 0
[R8-bgp]quit
[R8]
```

③ 过滤中国移动对外通告私网<10.0.0.0>私用网段

```
[R5]ip ip-prefix deny_net_10 deny 10.0.0.0 8 greater-equal 8
[R5]ip ip-prefix deny_net_10 index 20 permit 0.0.0.0 0 less-equal 32
[R5]bgp 200
[R5-bgp]filter-policy ip-prefix deny_net_10 export
[R5-bgp]quit
[R5]
```

(4) 联通网络 BGP 配置。

① 配置联通内网底层 IS-IS 路由，实现 LoopBack 接口之间连通。

请读者按照拓扑自行配置，以下为具体参数。

- R11 配置参数。

网络实体名称：20.0001.0000.0000.0001.00。

IS-IS 通告接口：GE0/0/0、GE0/0/1、S2/0/0、S2/0/1、LoopBack 0。

- R12 配置参数。

网络实体名称：20.0001.0000.0000.0010.00。

IS-IS 通告接口：GE0/0/0、GE0/0/1、LoopBack 0。

- R13 配置参数。

网络实体名称：20.0001.0000.0000.0011.00。

IS-IS 通告接口：GE0/0/0、GE0/0/1、S2/0/0、LoopBack 0。

- R14 配置参数。

网络实体名称：20.0001.0000.0000.0100.00。

IS-IS 通告接口：GE0/0/0、GE0/0/1、LoopBack 0。

② 请读者按照拓扑自行配置 BGP 协议实现联通和电信、移动网络互联。

路由器 R11 连接电信和移动网络，属于区域边界路由器。路由器 R12、R13 和 R14 分别与 R11 建立 IBGP 邻接关系，以学习到其他运营商内网路由。

③ 过滤中国电信对外通告私网<10.0.0.0>私用网段。

请读者按照拓扑自行配置。

(5) 在 R9、R10 和 R15 配置 Easy-IP 和静态路由。

```
[R9]acl 2000
[R9-acl-basic-2000]rule permit source any
[R9-acl-basic-2000]quit
[R9]interface Serial 2/0/0
```

```
[R9-Serial2/0/0]nat outbound 2000
[R9-Serial2/0/0]quit
[R9]ip route-static 0.0.0.0 0.0.0.0 114.32.32.1
```

请读者根据网络拓扑自行配置路由器 R10 和 R11 的 Easy-IP 和静态路由。

(6) 公司内网 Vlan 划分和 IGP 路由配置。

公司 A 内网配置。

```
[SW1]vlan batch 10 20
[SW1]port-group 1
[SW1-port-group-1]group-member GigabitEthernet 0/0/1 to GigabitEthernet 0/0/10
[SW1-port-group-1]port link-type access
[SW1-port-group-1]port default vlan 10
[SW1-port-group-1]quit
[SW1]port-group 2
[SW1-port-group-2]group-member GigabitEthernet 0/0/11 to GigabitEthernet 0/0/20
[SW1-port-group-2]port link-type access
[SW1-port-group-2]port default vlan 20
[SW1-port-group-2]quit
[SW1]interface GigabitEthernet 0/0/24
[SW1-GigabitEthernet0/0/24]port link-type trunk
[SW1-GigabitEthernet0/0/24]port trunk allow-pass vlan all
[SW1-GigabitEthernet0/0/24]quit
[SW1]interface Vlanif 1
[SW1-Vlanif1]ip address 192.168.1.1 24
[SW1-Vlanif1]quit
[SW1]interface Vlanif 10
[SW1-Vlanif10]ip address 192.168.10.1 24
[SW1-Vlanif10]quit
[SW1]interface Vlanif 20
[SW1-Vlanif20]ip address 192.168.20.1 24
[SW1-Vlanif20]quit
[SW1]isis 1
[SW1-isis-1]network-entity 49.0001.0000.0000.0001.00
[SW1-isis-1]is-level level-1
[SW1-isis-1]quit
[SW1]interface Vlanif 1
[SW1-Vlanif1]isis enable 1
[SW1-Vlanif1]interface Vlanif 10
[SW1-Vlanif10]isis enable 1
[SW1-Vlanif10]interface Vlanif 20
[SW1-Vlanif20]isis enable 1
[SW1-Vlanif20]quit
[SW1]ip route-static 0.0.0.0 0.0.0.0 192.168.1.2
[SW1]
------------------------------------------------------------------------------
[R9]isis 1
[R9-isis-1]network-entity 49.0001.0000.0000.0010.00
[R9-isis-1]is-level level-1
[R9-isis-1]quit
[R9]interface GigabitEthernet 0/0/0
[R9-GigabitEthernet0/0/0]isis enable 1
```

```
[R9-GigabitEthernet0/0/0]quit
[R9]
```

请读者根据网络拓扑，继续配置公司 B 和公司 C 内网 Vlan 和 IGP 路由。

2. 任务验证

(1) 查看边界路由器 BGP 对等体建立情况。

```
[R3]display bgp peer
 BGP local router ID : 202.202.202.2
 Local AS number : 100
 Total number of peers : 5     Peers in established state : 5
  Peer            V     AS  MsgRcvd  MsgSent  OutQ  Up/Down       State Pref  Rcv
  10.0.1.1        4    100        2       16     0  00:00:59     Established  0
  10.0.2.2        4    100        3       18     0  00:01:29     Established  0
  10.0.4.4        4    100        3       18     0  00:01:29     Established  0
  115.64.64.2     4    300       20       19     0  00:02:01     Established  12
  116.64.64.2     4    200       20       19     0  00:02:01     Established  12
```

可以看到，路由器 R3 分别与 R1、R2 和 R3 建立 IBGP 对等体，与 R5、R11 建立 EBGP 对等体，均为 Established 状态。

```
[R5]display bgp peer
BGP local router ID : 205.205.205.1
Local AS number : 200
Total number of peers : 5     Peers in established state : 5
  Peer            V     AS  MsgRcvd  MsgSent  OutQ  Up/Down       State Pref  Rcv
  10.0.2.2        4    200        6       20     0  00:04:43     Established  0
  10.0.3.3        4    200        6       20     0  00:04:30     Established  0
  10.0.4.4        4    200        6       19     0  00:04:46     Established  0
  116.64.64.1     4    100       22       24     0  00:05:15     Established  12
  117.64.64.2     4    300       23       23     0  00:05:18     Established  13
```

路由器 R5 分别与 R6、R6 和 R8 建立 IBGP 对等体，与 R3、R11 建立 EBGP 对等体，均为 Established 状态。

```
[R11]display bgp peer
BGP local router ID : 209.209.209.1
Local AS number : 300
Total number of peers : 5     Peers in established state : 5
  Peer            V     AS  MsgRcvd  MsgSent  OutQ  Up/Down       State Pref  Rcv
  10.0.2.2        4    300       11       27     0  00:09:04     Established  0
  10.0.3.3        4    300       11       27     0  00:09:04     Established  0
  10.0.4.4        4    300       11       27     0  00:09:04     Established  0
  115.64.64.1     4    100       26       28     0  00:09:19     Established  13
  117.64.64.1     4    200       27       28     0  00:09:22     Established  13
```

路由器 R11 分别与 R12、R13 和 R14 建立 IBGP 对等体，与 R3、R5 建立 EBGP 对等体，均为 Established 状态。

(2) 查看内部路由器 IP 路由表。

以 R1 为例，其路由表如下：

```
[R1]display ip routing-table                          //因篇幅限制,以下不列出直连网段路由条目
Route Flags: R - relay, D - download to fib
------------------------------------------------------------------------------
Routing Tables: Public
         Destinations : 35        Routes : 35
  Destination/Mask  Proto    Pre  Cost    Flags NextHop         Interface
        10.0.2.0/24 ISIS- L1  15  30        D   204.204.204.1   GigabitEthernet0/0/1
        10.0.3.0/24 ISIS- L1  15  20        D   204.204.204.1   GigabitEthernet0/0/1
        10.0.4.0/24 ISIS- L1  15  10        D   204.204.204.1   GigabitEthernet0/0/1
     115.64.64.0/24 ISIS- L1  15  30        D   204.204.204.1   GigabitEthernet0/0/1
     116.64.64.0/24 ISIS- L1  15  30        D   204.204.204.1   GigabitEthernet0/0/1
     117.64.64.0/24 IBGP     255  0         RD  116.64.64.2     GigabitEthernet0/0/1
     118.32.32.0/24 IBGP     255  50        RD  116.64.64.2     GigabitEthernet0/0/1
     119.32.32.0/24 IBGP     255  30        RD  115.64.64.2     GigabitEthernet0/0/1
   202.202.202.0/24 ISIS- L1  15  30        D   204.204.204.1   GigabitEthernet0/0/1
   203.203.203.0/24 ISIS- L1  15  20        D   204.204.204.1   GigabitEthernet0/0/1
   205.205.205.0/24 IBGP     255  0         RD  116.64.64.2     GigabitEthernet0/0/1
   206.206.206.0/24 IBGP     255  2         RD  116.64.64.2     GigabitEthernet0/0/1
   207.207.207.0/24 IBGP     255  2         RD  116.64.64.2     GigabitEthernet0/0/1
   208.208.208.0/24 IBGP     255  0         RD  116.64.64.2     GigabitEthernet0/0/1
   209.209.209.0/24 IBGP     255  0         RD  115.64.64.2     GigabitEthernet0/0/1
   210.210.210.0/24 IBGP     255  20        RD  115.64.64.2     GigabitEthernet0/0/1
   211.211.211.0/24 IBGP     255  20        RD  115.64.64.2     GigabitEthernet0/0/1
   212.212.212.0/24 IBGP     255  0         RD  115.64.64.2     GigabitEthernet0/0/1
```

可以看到,R1有去往公网所有网段路由条目。

(3) 连通性验证。

主机1～主机6可以访问公网任意网段。以主机1为例,与公网连通性情况如图14-2所示。

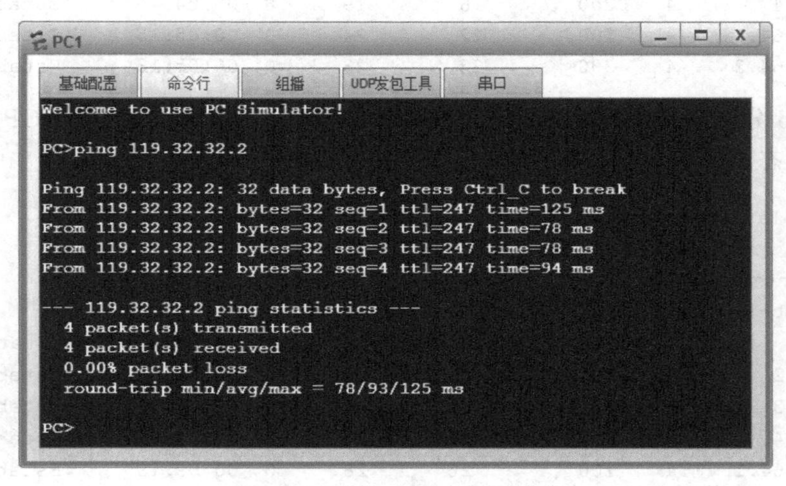

图14-2 主机1可以连通公网

附录

eNSP 使用技巧

1. 查看当前配置信息

```
<Huawei>display current-configuration          //可在任何视图和状态下查看
```

2. 取消操作配置(undo)

```
[Huawei-GigabitEthernet0/0/0]ip address 192.168.1.1 24
[Huawei-GigabitEthernet0/0/0]undo ip address
[Huawei]ospf 1
[Huawei-ospf-1]quit
[Huawei]undo ospf 1
Warning: The OSPF process will be deleted. Continue?[Y/N]:y
[Huawei]
```

3. 关闭系统信息提示

在使用华为 eNSP 模拟器对网络设备进行配置时，在执行某条命令后，经常会弹出信息提示。这些信息有的表示命令执行，相关配置已经生效；有的表示协议已启动及协议启动过程中状态变化情况。这些弹出信息提示会打断用户正在输入的命令，造成不便。此时可以在系统视图下输入 undo info-center enable 命令。

```
[Huawei]undo info-center enable
Info: Information center is disabled.          //系统提示:信息中心已无效
```

配置完成后，如果想再次启用信息提示，可以在系统视图下输入 info-center enable 命令。

```
[Huawei]info-center enable
Info: Information center is enabled.           //系统提示:信息中心已生效
```

注：初学者在配置过程中，不建议关闭信息提示，因为它能准确地反映输入配置命令是否生效和网络协议运行情况，提高配置准确率和工作效率。

4. 设置系统超时时长

使用华为 eNSP 模拟器对网络设备进行配置时，用户如在一段时间内没有对该设备进行任何操作，系统会自动退出到配置控制台视图，这段时间间隔称为空闲时长或设备连接超时时长。如果想继续配置，只能重新进入用户视图。空闲时长的设定在一定程度上能起到保护网络设备安全作用，但设置时长过短会给配置工作带来一些困扰。华为设备默认空闲时长为 10 分钟，如要进行调整，命令如下：

```
[Huawei]user-interface console 0          //由系统视图进入控制台视图
[Huawei-ui-console0]idle-timeout 30 0     //设置空间时长为 30 分钟。其中第一个参数表示分
                                            钟,第二个参数表示秒。分钟数默认值为 10,秒数
                                            默认值为 0
```

如果想将设备设置为永不超时，脚本如下：

```
[Huawei]idle-timeout 0                                    //注意，会导致安全问题
```

注：空闲时常需根据具体实施环境设置，务必确保设备配置安全。

5. 保存设备配置信息

华为 eNSP 模拟器窗口工具栏的“保存”按钮只用于保存网络拓扑和标识，如要保存设备配置信息，需在用户模式下输入 save 命令。

```
<Huawei>save
The current configuration will be written to the device.
Are you sure to continue?[Y/N]y
Save the configuration successfully.
<Huawei>
```

保存设备配置信息后，再单击工具栏中的“保存”按钮。

6. 清空设备配置信息并重新启动

```
<Huawei>reset saved-configuration                       //清空设备配置信息
This will delete the configuration in the flash memory.
The device configuratio
ns will be erased to reconfigure.
Are you sure? (y/n)[n]:y                                //选择 y，确定清空
 The config file does not exist.
<Huawei>reboot
Info: The system is comparing the configuration, please wait.
Warning: All the configuration will be saved to the next startup configuration.
Continue ?[y/n]:n   //是否将当前配置保存在启动配置文件中？选择 n，否则重启后原配置信息
                      还在
System will reboot! Continue ?[y/n]:y                   //是否继续重启，选择 y
Info: system is rebooting ,please wait...
<Huawei>
```

7. eNSP 连接虚拟机

云设备可以将任意设备连接在一起，如可通过云设备将虚拟机与 eNSP 的路由器 R1 相连，如附图 1 所示，步骤如下。

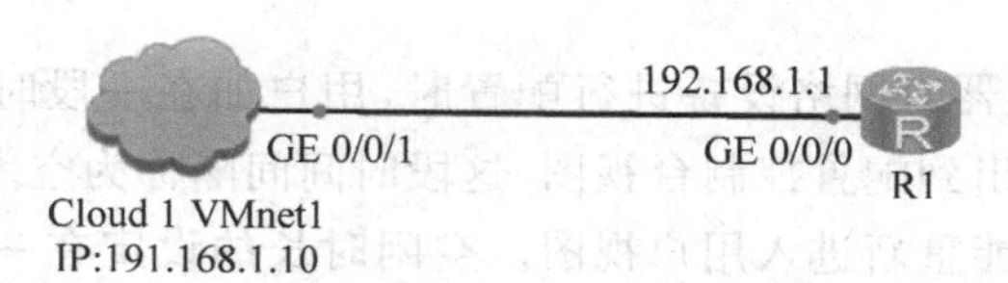

附图 1　虚拟机与路由器 R1 连接

(1) 选择云设备并拖动至工作区，双击 Cloud 1 进入配置界面。在“端口创建”选项区，将 UDP(61730 是动态端口号，由系统自动分配)与端口类型 GE 绑定(即在 Cloud 1 中创建 GE 接口，该接口通过 UDP 61730 端口与 R1 连接。也可以创建 Ethernet 接口，但速度较慢)，单击“增加”按钮添加绑定关系，如附图 2 所示。

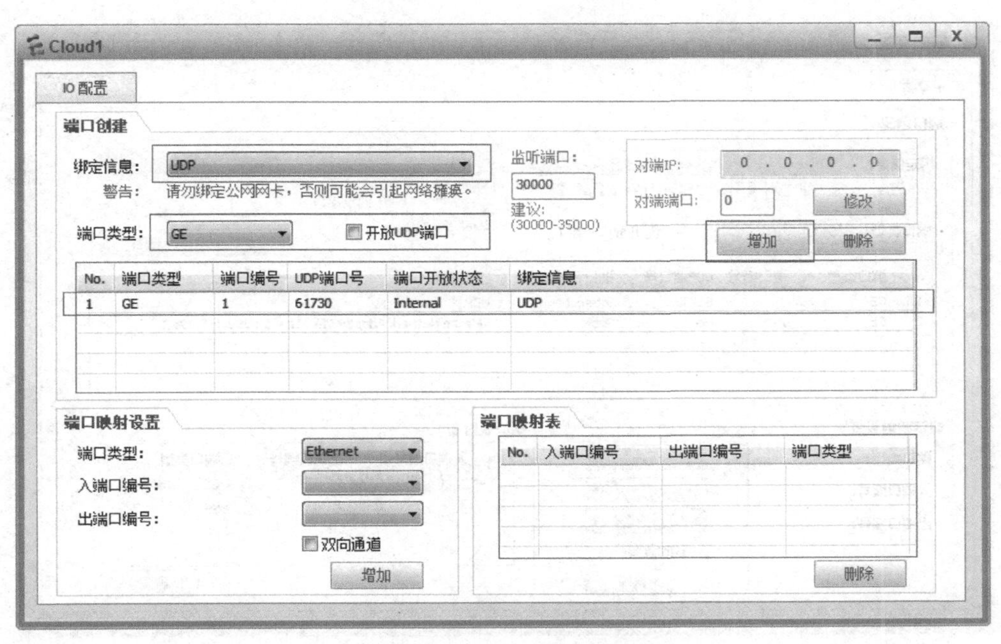

附图 2　绑定 UDP 与端口类型 GE

(2) 将 VMnet1 与端口类型 GE 绑定。单击“增加”按钮添加绑定关系，如附图 3 所示。此时 GE 充当中介作用，既通过 UDP 端口 61730 与路由器 R1 相连，又连接虚拟机 VMnet1 网络。

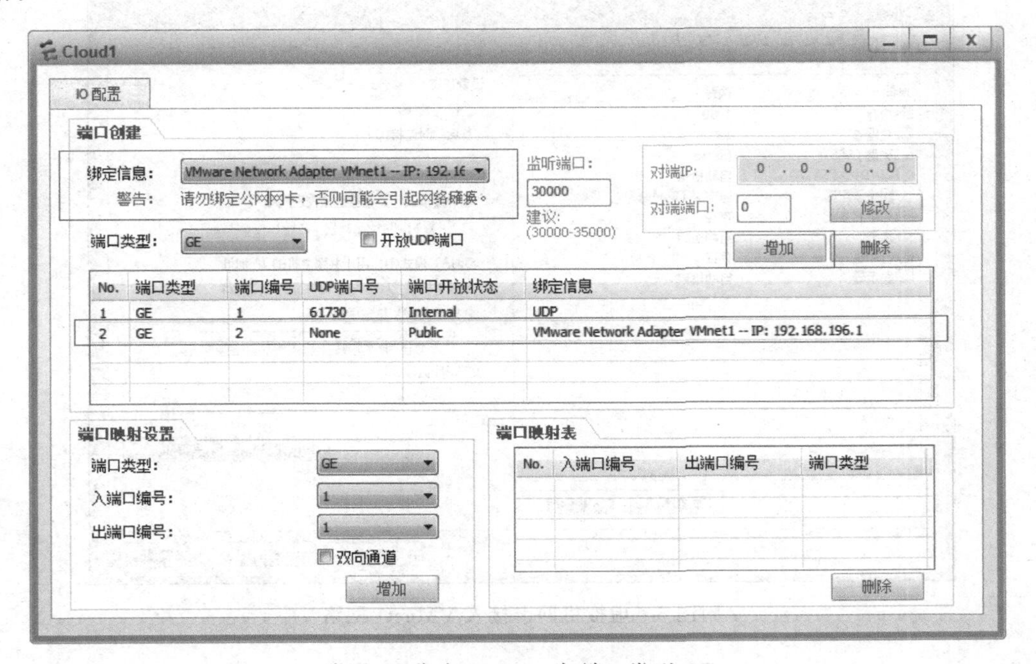

附图 3　绑定 VMnet1 与端口类型 GE

(3) 在“端口映射设置”中，添加 GE 入栈端口编号 1 与出栈端口编号 2，选中“双向通道”复选按钮，单击“增加”按钮添加至端口映射表，如附图 4 所示。

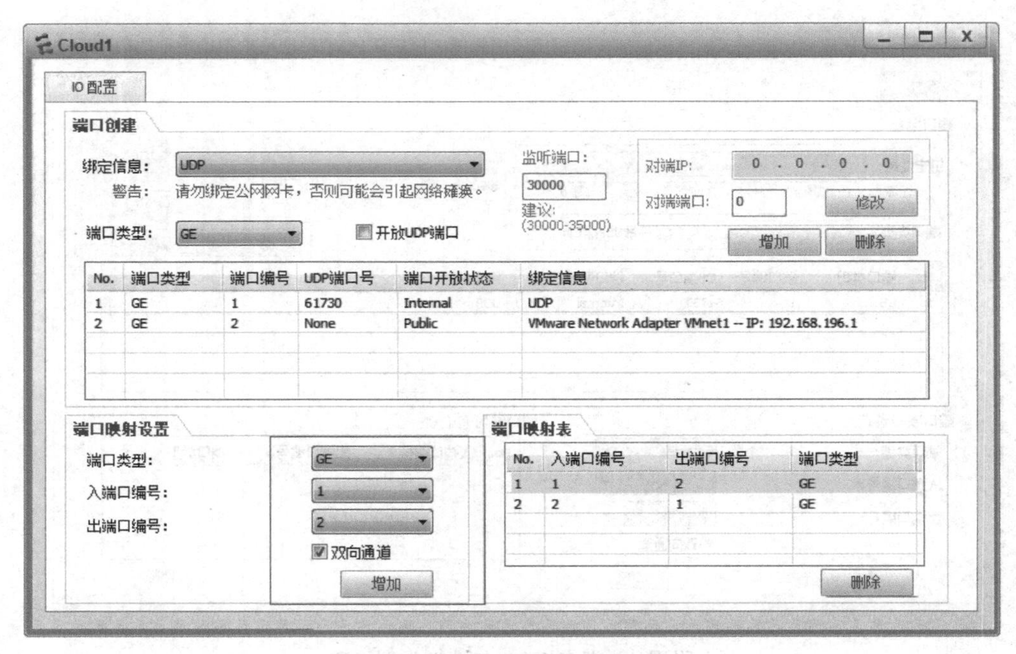

附图 4　添加通道端口映射表

(4) 将虚拟机系统网卡接入至 VMnet1 网络，如附图 5 所示，配置 IP 地址 192.168.1.10 后，与路由器 GE0/0/0 接口 IP(192.168.1.1)ping 通。

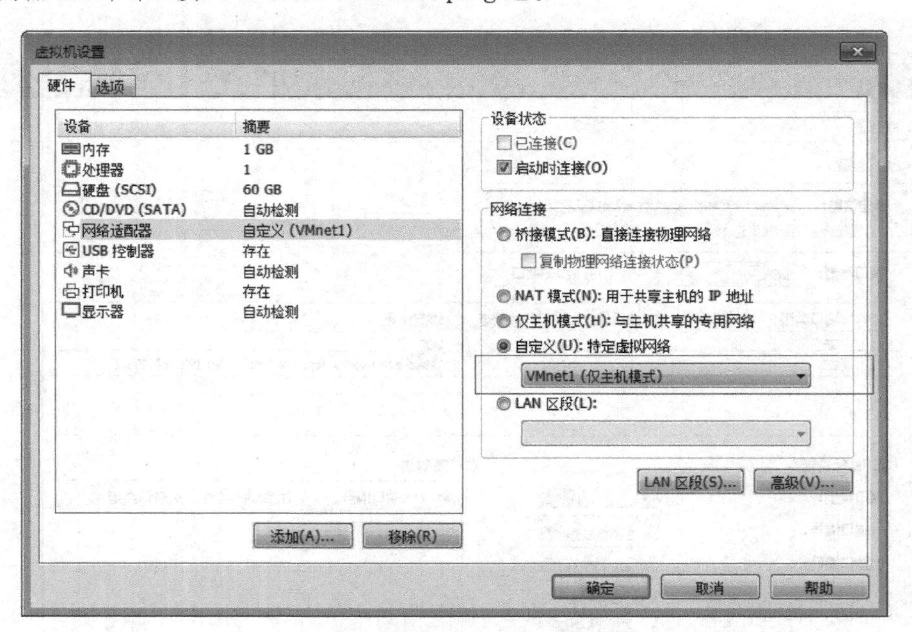

附图 5　虚拟机网卡接入 VMnet1 网络

注：eNSP 在开启状态下，假如计算机长时间进入休眠状态，唤醒后会发现虚拟机无法再次 ping 通路由器 R1 的情况。这是由于 GE 与路由器 R1 的 UDP 会话连接超时被断开，可在“端口创建”选项区，删除已绑定的 VMnet1 接口或者 UDP 端口(删除 UDP 端口后需要重新

接线)，再重新创建并绑定端口映射即可。

8. eNSP 设备连接 Internet

假如路由器 R1 需要连接至公网，如附图 6 所示，可按如下步骤操作。

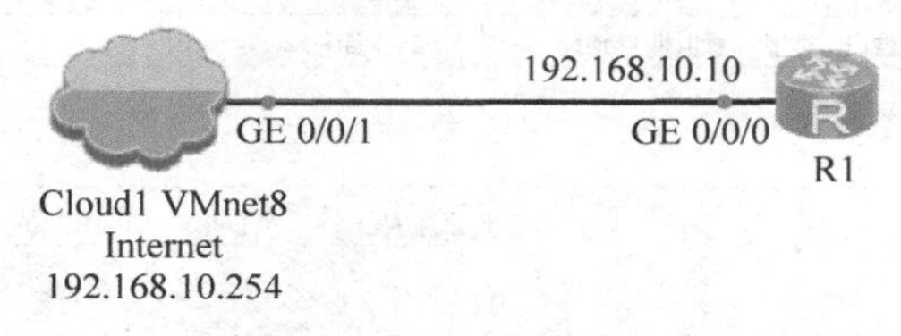

附图 6 将 R1 连接至公网

(1) 依次选择"虚拟机"→"编辑"→"虚拟网络编辑器"选项，打开"虚拟网络编辑器"窗口，将 VMnet8(NAT 模式)更改为＜192. 168. 10. 0＞网段，并单击"应用"按钮，如附图 7 所示。

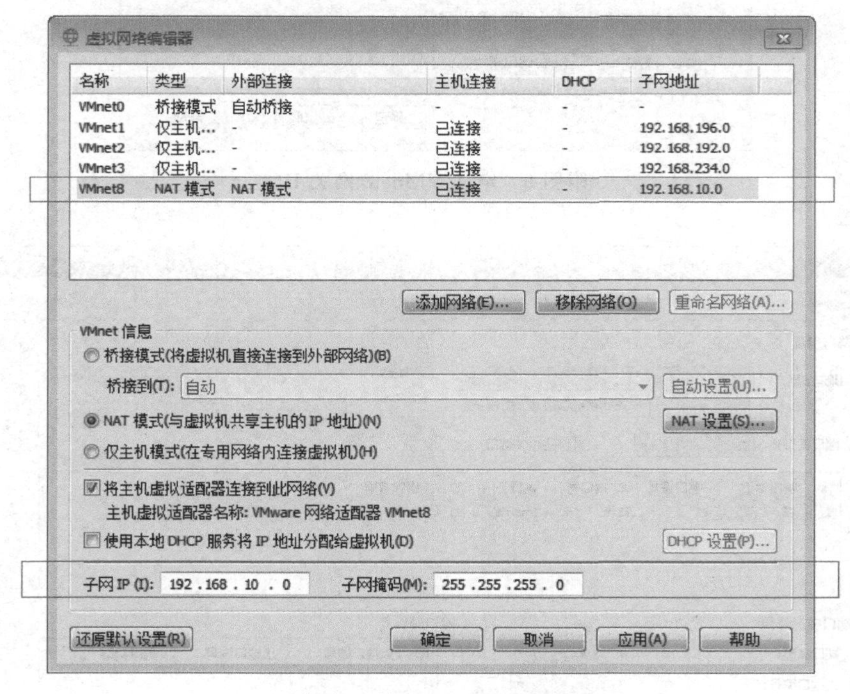

附图 7 定义 VMnet8 网段

(2) 单击"NAT 设置"按钮，定义 VMnet8 网关 IP 地址，如 192. 168. 10. 254，如附图 8 所示。

注：IP 地址 192. 168. 10. 1 默认已被 VMnet8 网卡占用，请定义其他 IP 地址作为网关 IP 地址。

(3) 选择云设备并拖动至工作区，双击 Cloud 1 进入配置界面，在"端口创建"选项区，将 UDP 与端口类型 GE 绑定，单击"增加"按钮添加绑定关系，如附图 9 所示。

(4) 将 VMnet8 与端口类型 GE 绑定，单击"增加"按钮添加绑定关系，如附图 10 所示。

(5) 在"端口映射设置"选项区，添加 GE 入栈端口编号 1 与出栈端口编号 2，选中"双向通道"复选框按钮，单击"增加"按钮添加至端口映射表。

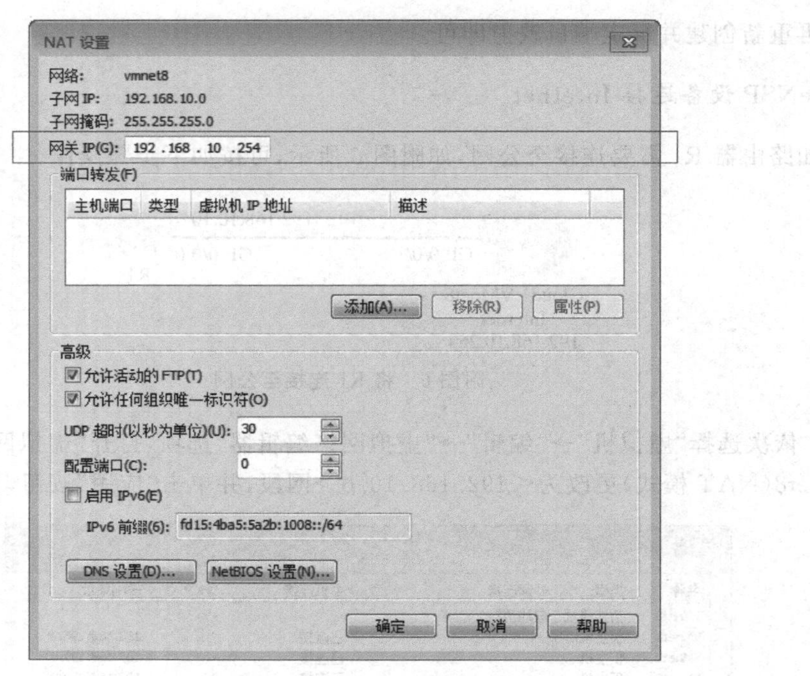

附图 8 定义 VMnet8 网关 IP

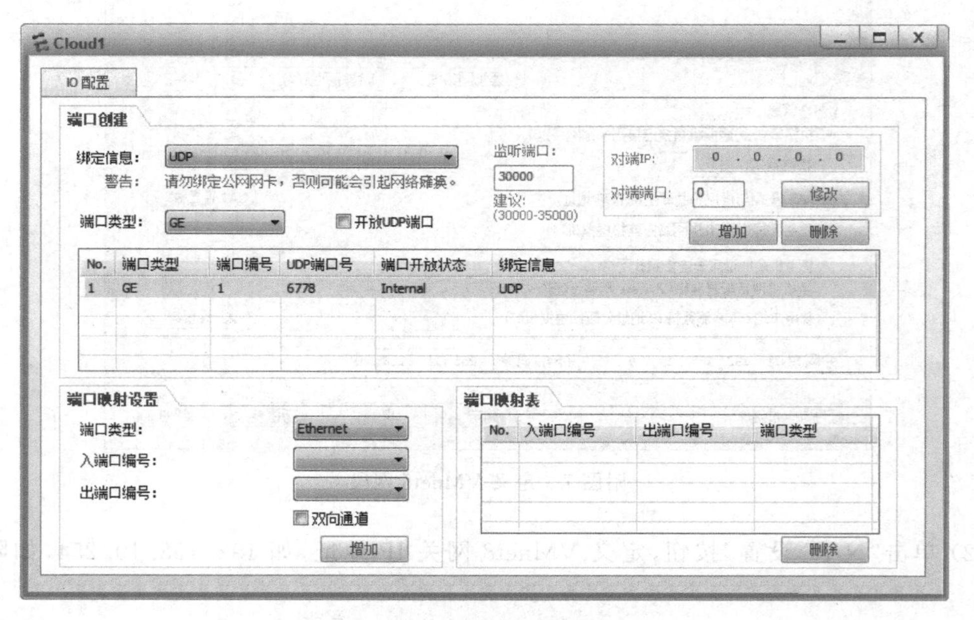

附图 9 绑定 UDP 与端口类型 GE

(6) 将路由器 R1 的 GE0/0/0 接口 IP 地址设置为 192.168.10.10，并配置默认路由，具体脚本如下：

```
[Huawei]interface GigabitEthernet 0/0/0
[Huawei-GigabitEthernet0/0/0]ip address 192.168.10.10 24
[Huawei-GigabitEthernet0/0/0]quit
[Huawei]ip route-static 0.0.0.0 0.0.0.0 192.168.10.254
```

附图 10　绑定 VMnet8 与端口类型 GE

(7) 在路由器 R1 中 ping 公网 IP,测试连通性。

```
[Huawei]ping 8.8.8.8
  PING 8.8.8.8: 56 Data bytes, press CTRL_C to break
    Reply from 8.8.8.8: bytes=56 Sequence=1 ttl=128 time=210 ms
    Reply from 8.8.8.8: bytes=56 Sequence=2 ttl=128 time=210 ms
    Reply from 8.8.8.8: bytes=56 Sequence=3 ttl=128 time=220 ms
    Reply from 8.8.8.8: bytes=56 Sequence=4 ttl=128 time=200 ms
    Reply from 8.8.8.8: bytes=56 Sequence=5 ttl=128 time=220 ms
  --- 8.8.8.8 ping statistics ---
    5 packet(s) transmitted
    5 packet(s) received
    0.00% packet loss
    round-trip min/avg/max = 200/212/220 ms
```

注:8.8.8.8 是公网 DNS 服务器地址,部署在美国。如内网 DNS 服务器故障,可在 DNS 中指定公网 DNS IP,如 8.8.8.8。

参 考 文 献

[1] 李锋. 网络设备配置与管理[M]. 北京：清华大学出版社，2020.
[2] 李锋. 网络技术基础与安全[M]. 北京：清华大学出版社，2014.
[3] 李锋. 基于华为 eNSP 网络攻防与安全实验教程[M]. 北京：清华大学出版社，2022.